AF667079

ZEITSCHRIFT FÜR ANGEWANDTE MATHEMATIK UND MECHANIK

INGENIEURWISSENSCHAFTLICHE FORSCHUNGSARBEITEN

UNTER MITWIRKUNG VON E. BECKER · H. BECKERT · L. BERG L. BITTNER · L. COLLATZ
W. FISZDON · H. GÖRTLER · J. HEINHOLD · K. MARGUERRE · P. H. MÜLLER · H. NEUBER
W. OLSZAK · K. OSWATITSCH · A. SAWCZUK · L. SCHMETTERER · G. SCHMIDT
K. SCHRÖDER · H. SCHUBERT · H. UNGER · C. WEBER UND F. WEIDENHAMMER
HERAUSGEGEBEN VON H. HEINRICH, DRESDEN

Vierundfünfzigster Jahrgang 1974

AKADEMIE-VERLAG · BERLIN

INHALTSÜBERSICHT

A. Verfasserverzeichnis

(B. = Bericht; Ber. = Berichtigung; H. = Hauptaufsatz; HV. = Hauptvortrag; KM. = Kleine Mitteilung; N. = Nachricht; V. = Vortragsauszug; ZB. = Zusammenfassender Bericht)

Der Buchstabe T vor der Seitenzahl weist darauf hin, daß der betreffende Beitrag im Sonderheft 4 (GAMM-Tagung 1973) erschienen ist.

B. Sachverzeichnis

(Abkürzungen s. S. II)

Dynamik, Schwingungen und Wellen

(siehe auch Elasto- und Plastomechanik; Mechanik, allgemein und Physik; Operationsforschung, Optimierung, optimale Prozesse; Stäbe, Schalen, Platten, Scheiben)

Eigenwertprobleme

(siehe auch Differential- und Integralgleichungen)

Elasto- und Plastomechanik

(siehe auch Methoden der finiten Elemente; Stäbe, Schalen, Platten, Scheiben; Wärmeübertragung)

Gasdynamik und Magnetohydrodynamik

(siehe auch Grenzschichttheorie und Turbulenz; Strömungsmechanik; Wärmeübertragung)

Grenzschichttheorie und Turbulenz

(siehe auch Gasdynamik und Magnetohydrodynamik; Wärmeübertragung)

Intervallmathematik

(siehe auch Numerische Mathematik und Rechentechnik)

Mathematik, allgemein

(siehe auch Approximationstheorie; Differential- und Integralgleichungen; Intervallmathematik; Matrizen; Numerische Mathematik und Rechentechnik; Operationsforschung, Optimierung, optimale Prozesse; Regelungssysteme; Wahrscheinlichkeitstheorie und Mathematische Statistik)

Matrizen

Mechanik, allgemein und Physik

(siehe auch Differential- und Integralgleichungen; Dynamik, Schwingungen und Wellen; Elasto- und Plastomechanik; Gasdynamik und Magnetohydrodynamik; Grenzschichttheorie und Turbulenz; Methode der finiten Elemente; Stäbe, Schalen, Scheiben; Strömungsmechanik; Wärmeübertragung)

Methode der finiten Elemente

Strömungsmechanik

(siehe auch Dynamik, Schwingungen und Wellen; Gasdynamik und Magnetohydrodynamik; Grenzschichttheorie und Turbulenz; Methode der finiten Elemente; Numerische Mathematik und Rechentechnik)

Wärmeübertragung

(siehe auch Elasto- und Plastomechanik; Gasdynamik und Magnetohydrodynamik; Numerische Mathematik und Rechentechnik)

Wahrscheinlichkeitstheorie und Mathematische Statistik

(siehe auch Numerische Mathematik und Rechentechnik)

Wissenschaftliche Gesellschaften, Tagungen, Nachrichten

C. Buchbesprechungen

(Der Name des Referenten ist in Klammern beigefügt)

Das Sonderheft 54—4 unterliegt nicht der Verantwortung des Herausgebers

Hinweise für Autoren

Die Zeitschrift für Angewandte Mathematik und Mechanik (ZAMM) veröffentlicht nur Originalbeiträge. Dissertationen, Habilitationsschriften, interne Forschungsberichte o. ä. sind als solche durch eine Fußnote zu kennzeichnen. Bei Einreichen des Manuskripts hat der Autor verbindlich zu erklären, daß sein Aufsatz noch in keiner anderen allgemein zugänglichen Zeitschrift veröffentlicht worden ist bzw. kein Antrag auf Veröffentlichung in einer solchen Zeitschrift läuft.

Manuskripte (im allgemeinen in einfacher, nur auf besondere Anforderung in mehrfacher Ausfertigung) und andere den Inhalt der Zeitschrift betreffende Zuschriften sind dem Herausgeber der ZAMM entweder über ein Mitglied des Herausgebergremiums oder direkt über folgende Anschrift zuzuleiten:

Zeitschrift für Angewandte Mathematik und Mechanik
z. Hd. Herrn Prof. Dr. Günter Schmidt
Zentralinstitut für Mathematik und Mechanik
an der Akademie der Wissenschaften der DDR
DDR-108 Berlin, Mohrenstraße 39

Als **Einreichtermin** gilt das Datum des Poststempels der Manuskriptsendung.

Die Manuskripte haben in ihrer äußeren Gestaltung folgenden Anforderungen zu genügen, anderenfalls sie zurückgewiesen werden können oder in der Bearbeitung lange Verzögerungen eintreten:

1. Die Manuskriptseiten sind nur einseitig zu beschreiben. Der Text ist mit Maschine mit weitem Zeilenabstand (mindestens $1\frac{1}{2}$-zeilig) und mit mindestens 3 cm breitem Rand zu schreiben.

2. Die Formeln müssen einwandfrei lesbar und mit großem Zeilenabstand geschrieben sein. Für mehrfach wiederkehrende umfangreiche oder drucktechnisch komplizierte Terme sind abkürzende Bezeichnungen einzuführen. Bei Exponentialausdrücken mit längerem Exponenten ist weitgehend von der Schreibweise $\exp(\cdots)$ Gebrauch zu machen. Die verwendeten Formelzeichen sind in einer für den Setzer eindeutigen Weise zu kennzeichnen und gegebenenfalls in einer besonderen Liste zu erläutern. Um unnötige Korrekturen zu vermeiden, ist im Manuskript besonders bei folgenden Buchstaben genau zu unterscheiden zwischen v, ν, γ und r, φ und Φ; ψ und Ψ; 0 (Null), O, o, Θ und θ; k und $\varkappa$; i und ι; ξ und ζ; x, X und χ; u und n. Die wichtigsten, insbesondere auch die im weiteren Verlauf des Textes zitierten Formeln sind am rechten Rand der Manuskriptseiten in runden Klammern durchzunumerieren.

3. Sätze, Hilfssätze, Korrolare, Definitionen, Beweise u. dgl. sind dadurch hervorzuheben, daß ihnen in Sperrdruck das betreffende Wort (Satz ..., Beweis..., usw.) vorangestellt wird. Die Aussage eines Satzes oder Hilfssatzes wird einheitlich kursiv gedruckt. Das Ende einer solchen Aussage oder eines Beweises ist im Manuskript zu markieren. Bei Hauptaufsätzen ist im Interesse einer übersichtlicheren Gliederung und besseren Lesbarkeit weitgehend von der Möglichkeit Gebrauch zu machen, einzelne Passagen (Beweise, Bemerkungen, ...) durch Kleindruck gegenüber dem Haupttext zurücktreten zu lassen und die Stellen durch den Hinweis „Petit" zu kennzeichnen.

4. Als Vorlagen für die **Abbildungen** sind saubere Zeichnungen in Bleistift oder Tusche (Beschriftung nur mit Bleistift!) auf gesonderten Blättern und in einem genügend großen, eine Verkleinerung gestattenden Format erforderlich.

5. Auf **Literatur** ist im Text durch Nummern in eckigen Klammern zu verweisen. Das nach diesen Nummern geordnete Literaturverzeichnis erscheint am Ende des Aufsatzes. Es wird gebeten, darin die Literatur nach folgendem Muster anzuführen:

bei Büchern:

Householder, A. S., The Theory of Matrices in Numerical Analysis, Blaisdell Publ. Comp., New York/Toronto/London 1964, p. 74.

bei Zeitschriftenartikeln:

Weissinger, J., Zur nichtlinearen Theorie der ungleichförmigen Umströmung von Profilen, ZAMM **50**, S. 337 bis 346 (1970).

Handelt es sich bei der zitierten Literaturstelle um eine Übersetzung, so ist anzugeben, wo bzw. in welcher Zeitschrift, in welchem Jahr und in welcher Sprache das Original erschienen ist.

6. Den Manuskripten ist auf gesondertem Blatt eine möglichst **kurze Zusammenfassung** in deutscher und/oder englischer und/oder russischer Sprache beizufügen. Falls sie nicht in allen drei Sprachen geliefert werden kann, ist wenigstens eine Liste der Fachausdrücke in der jeweils fehlenden Sprache erwünscht.

7. Dem Manuskript ist die **Anschrift des Autors** beizufügen, die am Ende des Aufsatzes zur Information des Lesers veröffentlicht wird. Darüber hinaus sind diejenigen Anschriften anzugeben, an welche die Reinzeichnungen und Druckfahnen zur Korrektur und die Sonderdrucke zu senden sind. Ändert sich während der Laufzeit einer Arbeit eine Anschrift, so ist dies zur Vermeidung von Fehlleitungen und langen Verzögerungen dem Herausgeber der ZAMM unverzüglich mitzuteilen. Die Autoren erhalten von den Hauptaufsätzen 75, von den Kleinen Mitteilungen 25 Sonderdrucke ohne Berechnung, darüber hinaus bis zu 250 Sonderdrucke gegen Berechnung.

To the attention of potential contributors

The "Zeitschrift für Angewandte Mathematik und Mechanik (ZAMM)" will only publish original papers. Doctoral theses, progress reports for internal use, etc., must be marked as such in a footnote, if they are to be published in ZAMM. On submitting manuscripts authors must give the explicit assurance that the paper submitted has not previously been published in, nor been submitted for publication to, another public journal.

Manuscripts (generally in one copy only, unless specifically requested otherwise) and other correspondence relating to the contents of the journal should be addressed to the editor of ZAMM, either through any one member of the editorial board or direct to the following address:

Zeitschrift für Angewandte Mathematik und Mechanik
c/o Prof. Dr. Günter Schmidt
Zentralinstitut für Mathematik und Mechanik
an der Akademie der Wissenschaften der DDR
DDR-108 Berlin, Mohrenstr. 39

The postmark will be considered as the **date of submission** for publication.

Manuscript must comply with the following style rules to avoid the risk of delayed publication or even rejection.

1. The text should be typewritten (use one side of the paper only) and adequately spaced (1½-line space, at least, between every two lines, margines not narrower than 3 cm).

2. Formulae must be clearly legible and should be spaced out sufficiently. Long terms that occur especially frequently or terms that present especial difficulty to type-setters should be abbreviated, if possible. In exponential experessions with comparatively long exponents the abbreviation "exp(· · ·)" should preferably be used throughout. Signs used in formulae should be clearly recognizable to type-setters or, if necessary, explained in an appended list. To avoid tiresome corrections special care should be taken to clearly distinguish the following letters in manuscripts: v, ν, γ and r; φ and Φ; ψ and Ψ; 0 (zero), O, o, Θ and θ; k and $\varkappa$; i and ι; ξ and ζ; x, X and x; u and n. Major formulae, in particular the ones quoted throughout the text, should be numbered continually on the r i g h t margin (in parentheses).

3. Theorems, corollaries, definitions, proofs, etc should be emphasized by printing, in spaced type, the relevant term (theorem . . ., proof . . ., etc.) in front of the expression, which, in its turn, will be printed uniformly in italics. The end of a theorem, proof. etc. should be marked in the manuscript. Certain passages (proofs, remarks, etc). in articles and monographs may be printed in smaller type to ensure easy readability and should, therefore, be marked "petit" in the manuscript.

4. **Illustrations** should be in the form of clear pencil or ink drawings (captions, legends, etc. in pencil only!) on s e p a r a t e sheets, big enough to be scaled down.

5. **References** should be numbered in brackets throughout the text, with the bibliography according to these numbers appended. References should follow the model printed below:

books:

HOUSEHOLDER, A. S., The Theory of Matrice sin Numerical Analysis, Blaisdell Publ. Comp., New York/Toronto/London 1964, p. 74.

articles:

WEISSINGER, J., Zur nichtlinearen Theorie der ungleichförmigen Umströmung von Profilen, ZAMM **50**, S. 337 bis 346 (1970).

In cases of translated references, the original source (name of periodical, year of publication, original language) should be stated.

6. **A concise summary** in German and/or English and/or Russian should be provided on a separate sheet. If a summary cannot be provided in all of the three languages, a list of technical terms in the missing language will be appreciated.

7. Authors should state their **address**, which will be published at the end of the contribution for the information of readers. Addresses should also be stated of persons to whom galley proofs and drawings and reprints are to be sent. The editors of ZAMM should at once be notified of any change of the author's address occuring after submission of a contribution, so as to avoid any postal errors and delays. Authors of major articles are allowed 75, and authors of other contribution 25, reprints gratis; they may get additional copies (250 at most) at cost.

Bestellungen sind an das zuständige Postamt oder eine Buchhandlung zu richten, Bestellunggen aus dem Ausland an die Firma Buchexport, Volkseigener Außenhandelsbetrieb der DDR, 701 Leipzig, Leninstr. 72, oder direkt an den Akademie-Verlag, DDR-108 Berlin, Leipziger Str. 3–4

Herausgeber und Hauptschriftleiter: Prof. Dr.-Ing. habil. Dr. techn. h. c. H. Heinrich, 8027 Dresden, Friedrich-Hegel-Str. 31. Verlag: Akademie-Verlag, DDR-108 Berlin, Leipziger Straße 3–4; Fernsprecher: 2200441, Telex-Nr. 112020, Postscheckkonto: Berlin 35021. Der vorliegende Jahrgang 54 (1974) der Zeitschrift für Angewandte Mathematik und Mechanik umfaßt 11 Hefte und ein Sonderheft. Bezugspreis: jährl. M 264,– (Sonderpreis für die DDR 216,–)
Veröffentlicht unter Lizenznummer 1282 des Presseamtes beim Vorsitzenden des Ministerrates der Deutschen Demokratischen Republik.
Gesamtherstellung: VEB Druckerei „Thomas Müntzer", 582 Bad Langensalza.
Printed in the German Democratic Republic.

ZEITSCHRIFT FÜR ANGEWANDTE MATHEMATIK UND MECHANIK

INGENIEURWISSENSCHAFTLICHE FORSCHUNGSARBEITEN

UNTER MITWIRKUNG VON E. BECKER · H. BECKERT · L. BERG · L. BITTNER · L. COLLATZ W. FISZDON · H. GÖRTLER · J. HEINHOLD · K. MARGUERRE · P. H. MÜLLER · H. NEUBER W. OLSZAK · K. OSWATITSCH · A. SAWCZUK · L. SCHMETTERER · G. SCHMIDT K. SCHRÖDER · H. SCHUBERT · H. UNGER · C. WEBER UND F. WEIDENHAMMER
HERAUSGEGEBEN VON H. HEINRICH, DRESDEN

BAND 54 1974 HEFT 1

AUS DEM INHALT:

AKADEMIE-VERLAG GMBH · BERLIN

ZAMM Bd. 54 Nr. 1 S. 1—72 Berlin, Januar 1974

M 18,—

INHALT

Hauptaufsätze Seite

Kleine Mitteilungen

Wir bitten, alle Manuskriptsendungen und andere Zuschriften an folgende Anschrift zu richten: Zeitschrift für Angewandte Mathematik und Mechanik, z. Hd. Prof. Dr.-Ing. Dr. h. c. H. Heinrich, 8027 Dresden, Friedrich-Hegel-Str. 31. Zu den Arbeiten, die als Hauptaufsätze bestimmt sind, erbitten wir auf gesondertem Blatt eine kurze Zusammenfassung des Inhalts, nach Möglichkeit in deutscher, englischer und russischer Sprache; falls die Übersetzungen nicht geliefert werden können, ist wenigstens die Angabe spezieller Fachausdrücke in den verschiedenen Sprachen erwünscht. Ausführliche Hinweise für die Autoren, um deren strikte Berücksichtigung gebeten wird, finden sich im Anschluß an das Inhaltsverzeichnis des Jahrganges **53** (1973).

Die Autoren erhalten von den Hauptaufsätzen 75, von den Kleinen Mitteilungen 25 Sonderdrucke ohne Berechnung, darüber hinaus bis zu 250 Sonderdrucke gegen Berechnung.

Der Verlag behält sich für alle Beiträge das Recht der Vervielältigung, Verbreitung und Übersetzung vor.

Von I. I. GICHMAN / A. W. SKOROCHOD

Stochastische Differentialgleichungen

(Übersetzung aus dem Russischen)

In deutscher Sprache herausgegeben von

HELGA BUNKE

(Mathematische Lehrbücher und Monographien II. Abteilung, Band XXV)

1971. VI, 338 Seiten — gr. 8° — Leinen 59,— M

Bestell-Nr. 761 288 0 (5772)

Das Buch schließt eine Lücke in der Literatur, in der eine derartig umfassende Darstellung der Theorie der stochastischen Differentialgleichungen bisher fehlt. Es bietet dem Wahrscheinlichkeitsrechner, der auf dem Gebiet der Markowschen Prozesse, insbesondere auf dem Gebiet der Diffusionsprozesse und ihren Verallgemeinerungen, arbeitet, wichtiges Rüstzeug, und es ist zu erwarten, daß viele Ergebnisse wegen des Zusammenhangs mit stochastischen dynamischen Systemen auch Eingang in Anwendungen finden wird.

Bestellungen durch eine Buchhandlung erbeten

AKADEMIE-VERLAG

DDR-108 Berlin, Leipziger Str. 3—4

ZAMM 54, 1 – 8 (1974)

R. VAN DER BORGHT / J. O. MURPHY / J. M. STEINER

A Theoretical Investigation of Finite Amplitude Thermal Convection in Non-Newtonian Fluids

Für den Fall sechseckigen Umrisses einer Konvektionszelle endlicher Länge werden die Grundgleichungen hergeleitet, die die Wärmekonvektion in einer nichtnewtonschen Flüssigkeit beschreiben. Ihre numerische Lösung zeigt im Falle zweier freier Ränder, daß die viskoelastischen Effekte die Nusselt-Zahl nur wenig ändern. Immerhin ist die Nusselt-Zahl bei gegebener Rayleigh-Zahl durchweg – wenn auch nur wenig – größer als bei einer newtonschen Flüssigkeit vergleichbarer Viskosität. Diese theoretischen Voraussagen stehen im Einklang mit neueren experimentellen Ergebnissen im Falle zweier starrer Ränder.

The fundamental system of equations describing finite amplitude steady convection with hexagonal planform in a non-Newtonian fluid have been derived. The numerical solution of this system, in the case of two free boundaries, reveals that there is little change in the Nusselt number due to visco-elastic effects. However, the Nusselt number at any given Rayleigh number is consistently higher, although only slightly, than that of a Newtonian fluid of comparable viscosity. These theoretical predictions are in line with the recent experimental findings in the case of two rigid boundaries.

Выводятся основные уравнения, описывающие конвективный теплообмен в неньютоновской жидкости для конвективных камер шестиугольного контура. Их численное решение для случая двух свободных кромок показывает, что вязкоупругие явления мало отражаются на изменении критерия Нуссельта. Однако критерий Нуссельта оказывается для заданного критерия Рэлея всегда большим, хотя и не немного, чем для ньютоновской жидкости сравнимой вязкости. Эти теоретические предсказывания согласовываются с новыми экспериментальными результатами для случая двух жёстких кромок,.

1. Introduction

In a recent paper LIANG and ACRIVOS [1] have reported on an experimental investigation on buoyancy driven convection in horizontal layers of non-NEWTONian fluids confined between rigid, parallel, conducting surfaces. The heat transfer characteristics of these fluids were found to be similar to those obtained for a viscous NEWTONian fluid with similar PRANDTL number for values of the RAYLEIGH number up to $5 \cdot 10^4$. However, the NUSSELT numbers at a given RAYLEIGH number were found to be slightly but consistently higher than those of a NEWTONian fluid of comparable viscosity. This effect was greatest at large RAYLEIGH number ($> 10^4$) but, as pointed out by the authors, the reported data for the polymer solutions used (at these high RAYLEIGH numbers) are subject to some uncertainty and this would exaggerate the difference between the corresponding NEWTONian and non-NEWTONian values of the NUSSELT number.

The purpose of the present paper is to present a theoretical investigation of finite amplitude steady convection in non-NEWTONian fluids having small relaxation times. The fundamental equations of the problem are derived, using a GLANSDORFF-PRIGOGINE [2], [3] technique, and are then integrated numerically. A comparison between the values obtained shows that non-NEWTONian effects do not introduce very marked changes in the NUSSELT number. As in the observations, the NUSSELT number for the visco-elastic fluid is consistently higher than the one for the corresponding NEWTONian fluid.

Since the problem is formulated in rectangular cartesian coordinates there is no need to distinguish covariant and contravariant tensorial quantities. The subscript notation and the usual indicial summation convention will be employed throughout the paper.

2. Basic Equations

The basic equations of the problem will be derived from the general variational evolutionary principle first introduced by PRIGOGINE and GLANSDORFF [2], [3]. This principle states that the actual flow evolves in such a way as to keep the generalized entropy production a minimum with respect to arbitrary variations from it.

A modified form suitable for finding stationary solutions, for a non-NEWTONian fluid, can be written as follows [4]:

$$\delta\Phi = \iiint dx\, dy\, dz \left\{\delta u_i \left[\varrho_0 u_j \frac{\partial u_i}{\partial x_j} + \frac{\partial p}{\partial x_i} - g\,\varrho\,\lambda_i - \frac{\partial p_{ij}}{\partial x_j}\right] + \frac{\delta T}{T}\left[\varrho_0 C_v u_j \frac{\partial T}{\partial x_j} - K \nabla^2 T\right]\right\}. \qquad (1)$$

In this expression the variables have their usual meaning but it should be kept in mind that $\lambda_i = (0, 0, -1)$ and that K is the coefficient of heat conduction of the fluid, assumed to be constant.

The variational principle, as written here, assumes that viscous dissipation can be neglected and that the BOUSSINESQ approximation holds; i.e. we neglect variations in the density except in the buoyancy term.

In this paper we shall investigate the behaviour of a non-NEWTONian fluid whose visco-elastic nature is described by the stress-relaxing MAXWELL constitutive relation [5], [6]

$$p_{ij} + \tau_0 \frac{\delta p_{ij}}{\delta t} = \mu\left(\frac{\partial u_i}{\partial x_j} + \frac{\partial u_j}{\partial x_i}\right) \qquad (2)$$

where p_{ij} is the extra-stress tensor obtained upon separating the isotropic pressure from the general stress tensor, τ_0 is the MAXWELL relaxation time, μ is the steady-state molecular viscosity, and the convected derivative $\delta/\delta t$ represents a time derivative in a coordinate system which moves and deforms with the material:

$$\frac{\delta p_{ij}}{\delta t} = \frac{\partial p_{ij}}{\partial t} + u_k \frac{\partial p_{ij}}{\partial x_k} - p_{ik} \frac{\partial u_j}{\partial x_k} - p_{jk} \frac{\partial u_i}{\partial x_k} \,. \tag{3}$$

This constitutive relation can be regarded as a special case of that proposed by OLDROYD [7], [8], and is sufficient to reveal the basic effects of visco-elasticity on thermal instability, particularly in view of the extremely low shear rates involved in a natural convection experiment. Moreover, as pointed out by PORTEOUS and DENN [5], this constant viscosity type constitutive relation characterizes the rheological properties of dilute and molten polymer solutions. For dilute polymer solutions, τ_0 is of the order $10^{-2}-10^{-3}$ sec.

Since the relaxation time of currently known visco-elastic fluids in which buoyancy-driven convection would occur under normal laboratory conditions is very small [1], [9], [10], we may approximate the viscous stress tensor for time-independent ($\partial/\partial t \equiv 0$) motion, by the following expression:

$$p_{ij} = \mu \left(\frac{\partial u_i}{\partial x_j} + \frac{\partial u_j}{\partial x_i}\right) - \tau_0 \mu \left[u_k \frac{\partial}{\partial x_k}\left(\frac{\partial u_i}{\partial x_j} + \frac{\partial u_j}{\partial x_i}\right) - \left(\frac{\partial u_i}{\partial x_k} + \frac{\partial u_k}{\partial x_i}\right)\frac{\partial u_j}{\partial x_k} - \left(\frac{\partial u_j}{\partial x_k} + \frac{\partial u_k}{\partial x_j}\right)\frac{\partial u_i}{\partial x_k}\right]. \tag{4}$$

Substituting (4) in (1) after using the continuity equation

$$\frac{\partial u_j}{\partial x_j} = 0 \tag{5}$$

we obtain the following expression for the variational principle:

$$\delta\Phi = \iiint dx\, dy\, dz \left\{\delta u_i \left[\varrho_0 u_j \frac{\partial u_i}{\partial x_j} + \frac{\partial p}{\partial x_i} - g \varrho \lambda_i - \mu \nabla^2 u_i\right] + \right.$$
$$\left. + \tau_0 \mu \left[u_k \frac{\partial}{\partial x_k}(\nabla^2 u_i) - \left(2 \frac{\partial u_j}{\partial x_k} \frac{\partial^2 u_i}{\partial x_j \partial x_k} + \frac{\partial u_i}{\partial x_k}(\nabla^2 u_k)\right)\right] + \frac{\delta T}{T}\left[\varrho_0 C_v u_j \frac{\partial T}{\partial x_j} - K \nabla^2 T\right]\right\}. \tag{6}$$

We now introduce the following expressions for the pressure, density, temperature and velocity:

$$p = p_0 + \Pi f \,, \qquad \varrho = \varrho_0 + P f \,, \qquad T = T_0 + F f \,, \tag{7), (8), (9}$$

$$u_i = \left(\frac{DW}{k^2}\frac{\partial f}{\partial x}\,,\ \frac{DW}{k^2}\frac{\partial f}{\partial y}\,,\ W f\right), \tag{10}$$

where p_0, Π, ϱ_0, P, T_0, F and W are functions of z to be determined and f is a function of x and y. The expression (10) for u_i will satisfy the continuity equation (5) if $f(x, y)$ satisfies the following partial differential equation:

$$\frac{\partial^2 f}{\partial x^2} + \frac{\partial^2 f}{\partial y^2} = -k^2 f \,. \tag{11}$$

This equation admits a number of solutions [11] depending on the shape of the convective cell. We shall consider in this paper the case of hexagonal cells for which

$$f = \left(\frac{2}{3}\right)^{1/2} \left\{\cos[k y] + \cos\left[k\left(\frac{\sqrt{3}}{2} x + \frac{y}{2}\right)\right] + \cos\left[k\left(\frac{\sqrt{3}}{2} x - \frac{y}{2}\right)\right]\right\} \tag{12}$$

where k is a horizontal wave-number determining the actual size of the cell.

If we introduce the expressions (7) to (10) in the variational principle (6) it is easily seen that

$$\delta\Phi = \iiint dx\, dy\, dz \left\{\frac{\delta(DW)}{k^2}\left[\frac{\partial f}{\partial x} M_x + \frac{\partial f}{\partial y} M_y\right] + \delta W f M_z + \frac{\delta T_0 + f\,\delta F}{T}\left[\varrho_0 C_v u_j \frac{\partial T}{\partial x_j} - K \nabla^2 T\right]\right\} \tag{13}$$

where

$$M_i = \varrho_0 u_j \frac{\partial u_i}{\partial x_j} + \frac{\partial p}{\partial x_i} - g \varrho \lambda_i - \mu \nabla^2 u_i + \tau_0 \mu \left\{u_k \frac{\partial}{\partial x_k}(\nabla^2 u_i) - \left[2 \frac{\partial u_j}{\partial x_k}\frac{\partial^2 u_i}{\partial x_j \partial x_k} + \frac{\partial u_i}{\partial x_k}(\nabla^2 u_k)\right]\right\} \tag{14}$$

and $D \equiv d/dz$.

We wish Φ to be stationary for arbitrary variations in W, T_0 and F respectively. The corresponding EULER-LAGRANGE equations will be

$$-\frac{1}{k^2} D \left\langle\frac{\partial f}{\partial x} M_x + \frac{\partial f}{\partial y} M_y\right\rangle + \langle f M_z\rangle = 0 \,, \tag{15}$$

$$\left\langle\varrho_0 C_v u_j \frac{\partial T}{\partial x_j} - K \nabla^2 T\right\rangle = 0 \tag{16}$$

and

$$\left\langle f\left\{\varrho_0\, C_v\, u_j \frac{\partial T}{\partial x_j} - K\, \nabla^2 T\right\}\right\rangle = 0 \tag{17}$$

where the angular brackets denote the horizontal averages

$$\langle\ \rangle \equiv A \int\int [\ ]\, dx\, dy \tag{18}$$

and the constant A has been chosen in such a way as to satisfy the normalizing condition

$$\langle f^2\rangle \equiv A \int\int f^2\, dx\, dy = 1\,. \tag{19}$$

The evaluation of the averages given in (15) to (17) is a lengthy but straightforward problem and we shall give here only the results of these calculations.

Let us introduce the notations

$$M_i^1 = \varrho_0\, u_j \frac{\partial u_i}{\partial x_j}, \qquad M_i^2 = \frac{\partial p}{\partial x_i}, \qquad M_i^3 = -\, g\, \varrho\, \lambda_i, \qquad M_i^4 = -\, \mu\, \nabla^2 u_i\,,$$
$$M_i^5 = \tau_0\, \mu \left\{u_k \frac{\partial}{\partial x_k}(\nabla^2 u_i) - \left[2 \frac{\partial u_j}{\partial x_k} \frac{\partial^2 u_i}{\partial x_j\, \partial x_k} + \frac{\partial u_i}{\partial x_k}(\nabla^2 u_k)\right]\right\} \tag{20}$$

and

$$Op(M_i^e) \;=\; -\, \frac{1}{k^2}\, D \left\langle \frac{\partial f}{\partial x} M_x^e + \frac{\partial f}{\partial y} M_y^e\right\rangle + \langle f\, M_z^e\rangle \tag{21}$$

We then obtain the following results:

(i) $$Op(M_i^1) \;=\; -\, \frac{\varrho_0\, C}{k^2}\{2\, DW\, D^2W + W\, D^3W - 3\, k^2\, W\, DW\} \tag{22}$$

where C is defined by

$$2\, C = \langle f^3\rangle\,. \tag{23}$$

This value of C is $1/\sqrt{6}$ for hexagonal cells and zero for rolls, square or rectangular ones.

In obtaining equation (22) we have made use of the following averages:

$$\left\langle \frac{\partial f}{\partial x} \frac{\partial f}{\partial y} \frac{\partial^2 f}{\partial x\, \partial y}\right\rangle = \frac{k^4\, C}{4}\,, \qquad \left\langle f\left[\left(\frac{\partial f}{\partial x}\right)^2 + \left(\frac{\partial f}{\partial y}\right)^2\right]\right\rangle = k^2\, C\,, \tag{24), (25}$$

(ii) $Op(M_i^2) = 0\,,$ (iii) $Op(M_i^3) = g\, P\,.$ (26), (27)

The above equations are supplemented by the equation of state

$$\varrho = \varrho_0 + \alpha\, \varrho_0\, (T_0 - T) \tag{28}$$

where α is the coefficient of volume expansion [12].

Introducing the notations (8) and (9)

$$P f = -\, \alpha\, \varrho_0\, F f\,. \tag{29}$$

Multiplying this equation by f and taking the horizontal average

$$P = -\, \alpha\, \varrho_0\, F\,. \tag{30}$$

Therefore

$$Op(M_i^3) \;=\; -\, \alpha\, \varrho_0\, g\, F \tag{31}$$

(iv) $$Op(M_i^4) \;=\; \frac{\mu}{k^2}(D^2 - k^2)^2\, W \tag{32}$$

where this time use has been made of the average

$$\left\langle \left(\frac{\partial f}{\partial x}\right)^2 + \left(\frac{\partial f}{\partial y}\right)^2\right\rangle = k^2\,. \tag{33}$$

(v) The evaluation of $Op(M_i^5)$ is extremely lengthy and we shall give here only the final result

$$Op(M_i^5) \;=\; -\, \frac{\tau_0\, \mu\, C}{k^2} \times$$
$$\times\{-\, 2\, k^2\, W\, D^3W + W\, D^5W - 6\, D^2W\, D^3W - DW\, D^4W + 3\, a^4\, W\, DW + 5\, a^2\, DW\, D^2W\} \tag{34}$$

In arriving at this result the following averages have been useful:

$$\left\langle f\left[\left(\frac{\partial^2 f}{\partial x^2}\right)^2 + 2\left(\frac{\partial^2 f}{\partial x\,\partial y}\right)^2 + \left(\frac{\partial^2 f}{\partial y^2}\right)^2\right]\right\rangle = \frac{k^4\,C}{2}, \tag{35}$$

$$\left\langle f\left\{\frac{\partial f}{\partial x}\left(\frac{\partial^3 f}{\partial x^3} + \frac{\partial^3 f}{\partial x\,\partial y^2}\right) + \frac{\partial f}{\partial y}\left(\frac{\partial^3 f}{\partial y^3} + \frac{\partial^3 f}{\partial x^2\,\partial y}\right)\right\}\right\rangle = -\,k^4\,C\,, \tag{36}$$

$$\left\langle\left(\frac{\partial f}{\partial x}\right)^2\left[\frac{\partial^4 f}{\partial x^4} + \frac{\partial^4 f}{\partial x^2\,\partial y^2}\right] + 2\,\frac{\partial f}{\partial x}\,\frac{\partial f}{\partial y}\left[\frac{\partial^4 f}{\partial y\,\partial x^3} + \frac{\partial^4 f}{\partial x\,\partial y^3}\right] + \left(\frac{\partial f}{\partial y}\right)^2\left[\frac{\partial^4 f}{\partial y^4} + \frac{\partial^4 f}{\partial x^2\,\partial y^2}\right]\right\rangle = \frac{-\,k^6\,C}{2}, \tag{37}$$

$$\left\langle\frac{\partial f}{\partial x}\left[\frac{\partial^2 f}{\partial x^2}\,\frac{\partial^3 f}{\partial x^3} + \frac{\partial^2 f}{\partial y^2}\,\frac{\partial^3 f}{\partial x\,\partial y^2} + 2\,\frac{\partial^2 f}{\partial x\,\partial y}\,\frac{\partial^3 f}{\partial x^2\,\partial y}\right] + \frac{\partial f}{\partial y}\left[\frac{\partial^2 f}{\partial y^2}\,\frac{\partial^3 f}{\partial y^3} + \frac{\partial^2 f}{\partial x^2}\,\frac{\partial^3 f}{\partial x^2\,\partial y} + 2\,\frac{\partial^2 f}{\partial x\,\partial y}\,\frac{\partial^3 f}{\partial x\,\partial y^2}\right]\right\rangle = \frac{k^6\,C}{4}. \tag{38}$$

Combining the results given in (22), (26), (31), (32) and (34) we find that equation (15) can be written

$$\begin{aligned}\mu\,(D^2 - k^2)^2\,W = {} & g\,\alpha\,k^2\,\varrho_0\,F + \varrho_0\,C\,\{2\,DW\,D^2W + W\,D^3W - 3\,k^2\,W\,DW\} + \\ & + \tau_0\,\mu\,C\,\{5\,k^2\,DW\,D^2W - 2\,k^2\,W\,D^3W + 3\,k^4\,W\,DW - DW\,D^4W + \\ & + W\,D^5W - 6\,D^2W\,D^3W\}\,.\end{aligned} \tag{39}$$

Applying the same method to equations (16) and (17) respectively it can be shown that

$$\varrho_0\,C_v D(FW) - K\,D^2T_0 = 0 \tag{40}$$

and

$$\frac{K}{\varrho_0\,C_v}(D^2 - k^2)\,F = W\,DT_0 + C\,\{F\,DW + 2\,W\,DF\}\,. \tag{41}$$

The equations can now be put in non-dimensional form by using the transformations

$$T \to (\varDelta T)\,T_0\,,\qquad F \to (\varDelta T)\,F\,,\qquad D \to \frac{1}{d}\,D\,,\qquad k^2 \to \frac{a^2}{d^2}\,,\qquad W \to \frac{\varkappa}{d}\,W\,, \tag{42}$$

where $\varDelta T$ = the temperature drop across the convective layer,
d = the thickness of the layer,
a = a dimensionless horizontal wave-number,
$\varkappa = \dfrac{K}{\varrho_0\,C_v}$ = the thermal diffusivity.

Introducing, in addition, the following parameters:

$$R = \frac{g\,\alpha\,d^3\,\varDelta T}{\varkappa\,\nu} = \text{the RAYLEIGH number},\qquad \sigma = \frac{\nu}{\varkappa} = \text{the PRANDTL number},$$
$$\nu = \frac{\mu}{\varrho_0} = \text{kinematic viscosity}, \tag{43}$$

it is possible to write the basic equations of our problem in the following form

$$\begin{aligned}(D^2 - a^2)^2\,W = {} & R\,a^2\,F + \frac{C}{\sigma}\{2\,DW\,D^2W + W\,D^3W - 3\,a^2\,W\,DW\} + \\ & + C\,\tau_1\,\{5\,a^2\,DW\,D^2W - 2\,a^2\,W\,D^3W + 3\,a^4\,W\,DW - DW\,D^4W + \\ & + W\,D^5W - 6\,D^2W\,D^3W\}\,,\end{aligned} \tag{44}$$

$$(D^2 - a^2)\,F = W\,DT_0 + C\,(F\,DW + 2\,W\,DF)\,, \tag{45}$$

$$D^2T_0 = D(FW) \tag{46}$$

where τ_1 is an elastic parameter defined as follows

$$\tau_1 = \frac{\tau_0\,\varkappa}{d^2}\,. \tag{47}$$

In the case of two free, isothermal and non-deformable boundaries this system of equations will have to be solved subject to the following boundary conditions:

$$T_0(0) = 0\,,\qquad T_0(1) = -\,1 \tag{48}$$

and

$$W = D^2W = F = 0 \qquad \text{at} \qquad z = 0 \qquad \text{and} \qquad z = 1\,. \tag{49}$$

An outline of the numerical method used will be found in the next paragraph.

3. Numerical Method and Discussion of Results

The non-linear BÉNARD problem with hexagonal planforms ($C \neq 0$) and rigid boundaries has been solved, in the case of NEWTONian fluids, by ROBERTS [11], using a quasi-linearization technique and the RUNGE-KUTTA routine. However, at RAYLEIGH number $R \approx 3\,R_c$ the method failed to converge. The use of a finite-difference-scheme method by GOUGH, SPIEGEL and TOOMRE [13] in their study of this problem, with several horizontal modes included, was successful to very high values of R. Solutions when the free boundary conditions apply have been obtained by MURPHY [14] to high RAYLEIGH numbers, using expansions in truncated FOURIER series.

In the present problem on BÉNARD convection in viscoelastic fluids ($C \neq 0$, $\tau_1 \neq 0$) the differential equation (44) for $W(z)$ now has additional non-linear terms up to the fifth order in W. In order to satisfy the boundary conditions (48) and (49) we represent W, F, and T_0 by the following FOURIER expansions

$$W(z) = \sum_{n=1}^{M} W_n \sin n\pi z\,, \tag{50}$$

$$F(z) = \sum_{n=1}^{M} f_n \sin n\pi z\,, \tag{51}$$

$$T_0(z) = \sum_{n=1}^{M} t_n \sin n\pi z - z \tag{52}$$

and substitution into equations (44), (45) and (46) leads to the following system of $3M$ non-linear algebraic equations:

$$\begin{aligned}(n^2+\alpha^2)\,\omega_n = \frac{R_1\alpha^2}{(n^2+\alpha^2)} f_n + \frac{1}{(n^2+\alpha^2)} \sum_{p=1,2,\dots}^{M} \omega_p \{ |n-p|\,[C_3 + p^2(C_1 + p^2 C_5) + \\ + (n-p)^2 (C_2 + p^2 C_4 + (n-p)^2 C_6)\,]\,\omega_{|n-p|} - \\ - (p+n)\,[C_3 + p^2(C_1 + p^2 C_5) + (n+p)^2(C_2 + p^2 C_4 + (n+p)^2 C_6)]\,\omega_{n+p}\}\,,\end{aligned} \tag{53}$$

$$\begin{aligned}(n^2+\alpha^2)\,f_n = \omega_n - \frac{\pi}{2} \sum_{p=1,2,\dots}^{M} p\,t_p\,[\omega_{n+p} + Y(n-p)\,\omega_{|n-p|}] + \\ + \frac{C\pi}{2} \sum_{p=1,2,\dots}^{M} \omega_p \{(2n+p) f_{n+p} + [n\,Y(p-n) - |n-p|]\,f_{|n-p|}\}\,,\end{aligned} \tag{54}$$

$$n\,t_n = \frac{\pi}{2} \sum_{p=1,2,\dots}^{M} \omega_p\,[f_{n+p} + Y(p-n)\,f_{|n-p|}]\,, \qquad n = 1, 2, 3, \dots, M \tag{55}$$

where

$$R_1 = \frac{R}{\pi^4}\,, \qquad \omega_n = \frac{W_n}{\pi^2}\,, \qquad \alpha = \frac{a}{\pi}\,, \qquad q = \frac{C\pi}{\sigma} \tag{56}$$

$$Y(x) = \begin{cases} 1 & x > 0 \\ 0 & x = 0\,, \\ -1 & x < 0 \end{cases} \tag{57}$$

$$\begin{aligned} C_1 &= -\frac{5}{2} C\pi^2 \tau_1 \alpha^2 - q\,, & C_2 &= C\pi^3\tau_1\alpha^2 - \frac{q}{2}\,, & C_3 &= \frac{3}{2} C\pi^3\tau_1\alpha^4 - \frac{3}{2}\alpha^2 q\,, \\ C_4 &= -3\,C\pi^3\tau_1\,, & C_5 &= -\frac{C}{2}\pi^3\tau_1\,, & C_6 &= \frac{C}{2}\pi^3\tau_1\,. \end{aligned} \tag{58}$$

All even and odd vertical modes in (50), (51) and (52) have to be retained as the solutions sought no longer have symmetry about $z = 1/2$, as in the case of the $C = 0$ type solutions. The total number of modes M, included in any particular solution, depended upon a test for the constancy of the NUSSELT number N over the layer from $z = 0$ to $z = 1$. From a physical point of view, when $a = O(1)$, minimum M is essentially determined by the boundary layer thickness. This corresponds to the numerical requirement that there should be sufficient "collocation points" within the smallest boundary layer. For example at $R = 5 \cdot 10^4$, when $\tau_1 = 10^{-3}$ and $a = \pi/\sqrt{2}$ it was necessary to take $M = 60$ and the resulting $3\,M$ equations were solved on an iterative basis using the generalized NEWTON-RAPHSON procedure with N being evaluated from the integral

$$N = 1 + \int_0^1 F\,W\,dz\,. \tag{59}$$

Table 1

	$C = 1/\sqrt{6}$, $\sigma = 10^2$					
	$R = 5 \cdot 10^3$		$R = 10^4$		$R = 5 \cdot 10^4$	
a/π	$\overline{N}(\tau_1 = 0)$	$N(\tau_1 = 10^{-3})$	$\overline{N}(\tau_1 = 0)$	$N(\tau_1 = 10^{-3})$	$\overline{N}(\tau_1 = 0)$	$N(\tau_1 = 10^{-3})$
0.25	2.2783	2.2800	3.1757	3.1790	5.8889	5.9254
0.5	2.5968	3.6021	4.7546	4.7689	8.6072	8.9030
0.707	4.0151	4.0234	5.3335	5.3570	9.7180	10.2729
1.0	4.0389	4.0488	5.4955	5.5229	10.2741	11.0697
1.25	3.7395	3.7492	5.2818	5.3073	10.2571	10.9254
1.5	3.2530	3.2626	4.8758	4.8982	10.0258	10.5228
1.75	2.6372	2.6454	4.3315	4.3511	9.6701	10.0568
2.0	1.9452	1.9519	3.6750	3.6920	9.2300	9.5375
2.25	1.2707	1.2718	2.9316	2.9458	8.7096	8.9572
2.50	—	—	2.1478	2.1593	8.1048	8.3039
2.75	—	—	1.4144	1.4201	7.4098	7.5680
3.0	—	—	—	—	6.6216	6.7448

In Table 1 the NUSSELT number $\overline{N}$ for a NEWTONian fluid, of PRANDTL number $\sigma = 10^2$ for hexagonal planform of the convective cells, can be compared with the values N obtained for a viscoelastic fluid for which the dimensionless relaxation time $\tau_1 = 10^{-3}$. This value of τ_1 is probably an upper limit for this parameter; it has been obtained in the case where the relaxation time of the fluid is 1 sec, the thermal diffusivity is 10^{-3} cm²/sec, and for a layer thickness of 1 cm.

As can be seen from the Table 1 there is only a slight change in the NUSSELT number at RAYLEIGH number $R \leqq 10^4$. At RAYLEIGH number $R = 5 \cdot 10^4$ the change is more marked and, in line with the experimental results, the NUSSELT number is consistently larger for a non-NEWTONian fluid than for the NEWTONian fluid with same PRANDTL number. This is consistent with the destabilizing effect of elasticity on thermal convection [9], [15]—[19].

In Table 2 we illustrate the dependence of the NUSSELT number on RAYLEIGH number for both a NEWTONian and non-NEWTONian fluid. These results were obtained for a wave-number $a_c = \pi/\sqrt{2}$ corresponding to the critical RAYLEIGH number R_c predicted by the linear theory. This wave-number was chosen on the assumption that the linear theory can make good predictions of the horizontal wave-number of the convective flow pattern; the reason for this being that convection in the range of the RAYLEIGH numbers considered is an organized steady flow which is dependent upon its past history [20], [21].

Table 2

	$C = 1/\sqrt{6}$, $\sigma = 10^2$, $a = \pi/\sqrt{2} = a_c$	
R	$\overline{N}(\tau_1 = 0)$	$N(\tau_1 = 10^{-3})$
10^3	1.5995	1.6006
$2 \cdot 10^3$	2.5830	2.5858
$4 \cdot 10^3$	3.6410	3.6472
$6 \cdot 10^3$	4.3372	4.3480
$8 \cdot 10^3$	4.8800	4.8966
10^4	5.3335	5.3570
$2 \cdot 10^4$	6.9547	7.0318
$4 \cdot 10^4$	8.9690	9.2785
$5 \cdot 10^4$	9.7180	10.2729

It can be seen that the values of the NUSSELT number for a visco-elastic fluid are again consistently higher, although only slightly, than the corresponding ones for a NEWTONian fluid with same PRANDTL number and wave-number. Moreover, this difference in the NUSSELT number becomes more marked with the increasing RAYLEIGH number. The dependence of N on R is illustrated in Figure 4 for free boundaries when the convective cells have hexagonal planform. It is of interest to compare the corresponding dependence obtained experimentally by LIANG and ACRIVOS [1] for various Sevaran AP-30 solutions in the case of rigid boundaries, bearing in mind that, for the same value of R, less heat will be transported when the fluid layer is contained by rigid boundaries [22] rather than by free boundaries in the case of a NEWTONian fluid.

The detailed solutions obtained for the vertical velocity $W(z)$, the temperature fluctuation $F(z)$ and the mean temperature $T_0(z)$ across the fluid layer $0 \leqq z \leqq 1$, for several values of R when $\tau_1 = 10^{-3}$ are given in Figure 1. For large PRANDTL number, $\sigma = 10^2$, and when $R \geqq 10^4$ these solutions show that the greater maximum in the temperature fluctuation exists near the upper boundary.

In Figure 2 the W, F and T_0 solutions are given for both $\tau_1 = 0$ and $\tau_1 = 10^{-3}$ at two values of R. This comparison shows that at $R = 10^4$ there is little difference between the solutions for the two types of fluid but at $R = 5 \cdot 10^4$ there are some marked differences due to the viscoelastic effects, as already noted in the values obtained for the total heat transport. In particular, we observe that larger vertical velocities are attained in the viscoelastic fluid and that the maximum is shifted nearer the upper boundary.

The dependence of the W, F and T_0 solutions on the horizontal wave number a is illustrated in figure 3 for the viscoelastic fluid when $R = 5 \cdot 10^4$. The corresponding values of N are given in Table 1. One different feature arising from the solutions obtained from the smaller convective cell size when $a = 3\pi$ is that the vertical velocity, apart from having a reduced maximum value, is near constant over the central regions of the fluid layer.

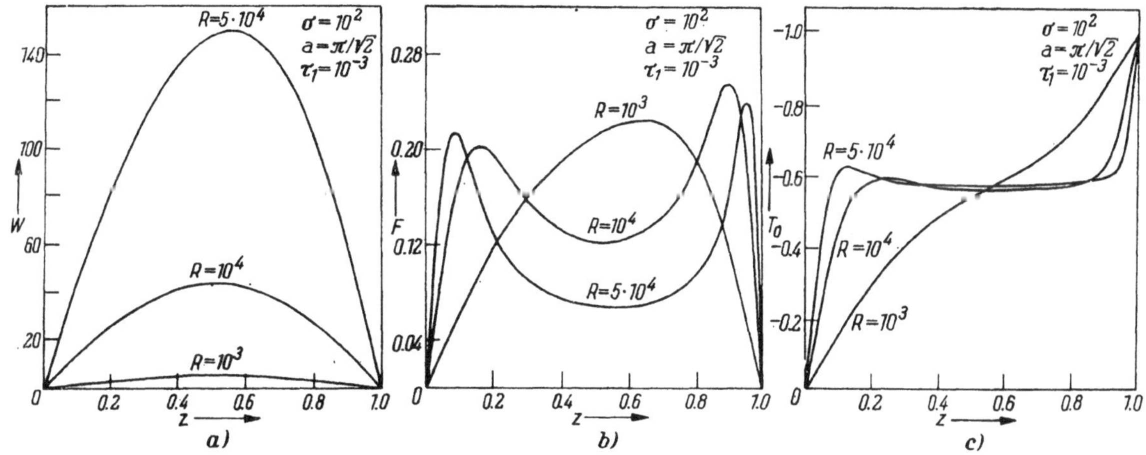

Fig. 1. Numerical solutions obtained for W, F and T_0 as functions of $z, 0 \leqq z \leqq 1$, when $a = \pi/\sqrt{2}$, $\sigma = 10^2$, $\tau_1 = 10^{-3}$ and for the indicated values of the RAYLEIGH number R

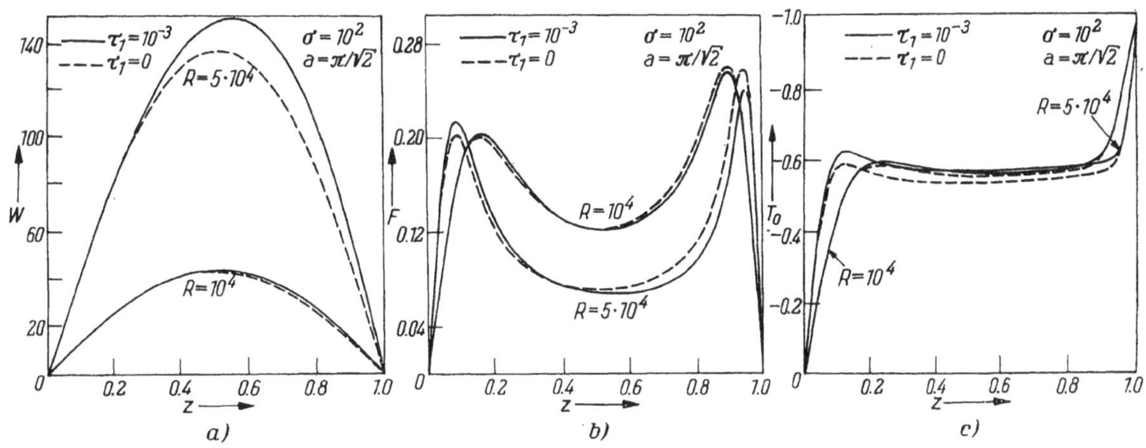

Fig. 2. Numerical solutions obtained for W, F and T_0 as functions of z, $0 \leqq z \leqq 1$, when $a = \pi/\sqrt{2}$, $\sigma = 10^2$, $\tau_1 = 10^{-3}$ (———), $\tau_1 = 0$ (--------) and for the two indicated values of R

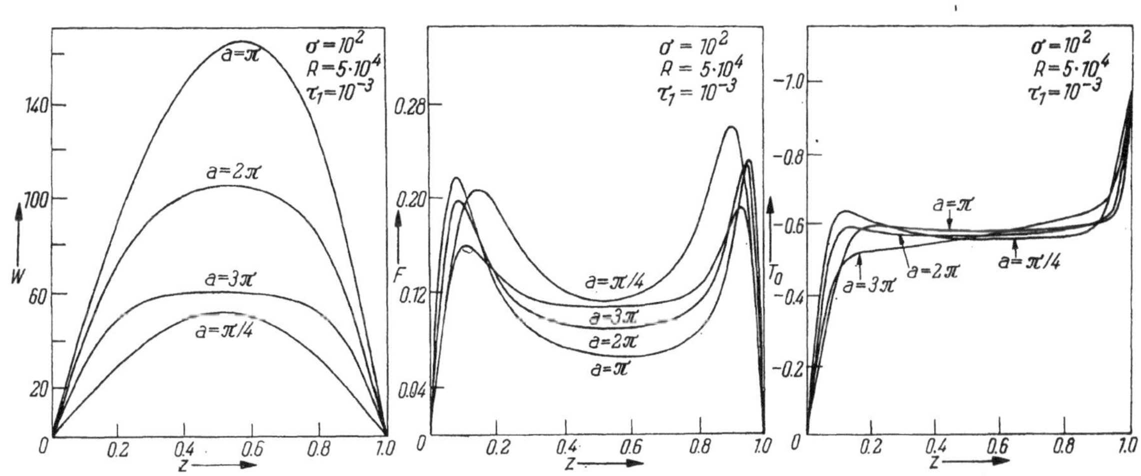

Fig. 3. Numerical solutions obtained for W, F and T_0 as functions of z, $0 \leqq z \leqq 1$ when $R = 5 \cdot 10^4$, $\sigma = 10^2$, $\tau_1 = 10^{-3}$ and for the indicated values of the wave number a

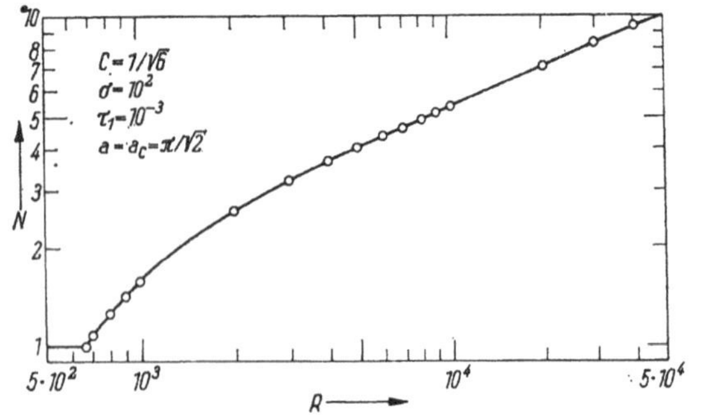

Fig. 4. Variation of the NUSSELT number N with the RAYLEIGH number R when $C = 1/\sqrt{6}$, $\sigma = 10^2$, $\tau_1 = 10^{-3}$ and $a = \pi/\sqrt{2}$

4. Concluding Remarks

The major result that emerges from the above theory is that when a MAXWELL fluid is subjected to buoyancy-driven convection, between two free isothermal and non-deformable boundaries, there is little change in the NUSSELT number due to visco-elastic effects. However, the NUSSELT number is consistently higher, although only slightly, than the one for an ordinary viscous fluid with same PRANDTL number. It would appear then from the above results, that when subject to buoyancy-driven steady convection, a visco-elastic fluid acts rather like an ordinary viscous fluid with constant comparable viscosity. Of course, one would expect a significant departure from such an approximate NEWTONian behaviour at higher rates of strain, however, this would probably require values of the RAYLEIGH number substantially larger than those covered during the course of this investigation.

These predictions are in good qualitative agreement with the recent experimental findings of LIANG and ACRIVOS [1] on buoyancy-driven convection in horizontal layers of dilute water solutions of polyacrylamides confined between rigid, parallel conducting surfaces. By determining the temperature difference at the point where convection sets in, they have calculated values of the zero shear rate viscosity which were in good agreement with those obtained from rheogoniometer data. The zero shear rate viscosity is one of the most important physical parameters of a non-NEWTONian fluid, hence it is important to determine it accurately. Unfortunately, it is not easy to achieve this task by conventional means (capillary viscometers, rheogoniometers). However, in a natural convection experiment the fluid velocities at the point at which convection first sets in are everywhere very small; hence, the viscosity entering into the expression for R_c should be that corresponding to zero strain. Using this fact, the authors have used for the first time a theoretically determined critical RAYLEIGH number in conjunction with a convection experiment to obtain the value of a physical quantity.

The numerical solution of the basic equations for the case of two rigid boundaries is thus clearly an obvious step in the development of the theory and we hope to explore this in the near future.

We are greatful to the referee for his valuable comments.

References

1 LIANG, S. F. and ACRIVOS, A., Rheologica Acta **9**, p. 447 (1970).
2 PRIGOGINE, I. and GLANSDORFF, P., Physica **30**, p. 351 (1964).
3 PRIGOGINE, I. and GLANSDORFF, P., Physica **31**, p. 1242 (1965).
4 UNNO, W., Publ. Astr. Soc. Japan **21**, p. 240 (1969).
5 PORTEOUS, K. C. and DENN, M. M., Linear und non-linear stability of plane Poiseuille flow of viscoelastic liquids, University of Delaware Water Resources Centre, Contribution No. 16 (1971).
6 TLAPA, G. and BERNSTEIN, B., Physics Fluids **13**, p. 565 (1970).
7 OLDROYD, J. G., Proc. Roy. Soc. London, A **200**, p. 523 (1950).
8 OLDROYD, J. G., Proc. Roy. Soc. London, A **245**, p. 278 (1958).
9 VEST, C. M. and ARPACI, V. S., J. Fluid Mech. **36**, p. 613 (1969).
10 GREEN, III, T., Physics Fluids **11**, p. 1410 (1968).
11 ROBERTS, P. H., Non-Equilibrium Thermodynamics, Variational Techniques and Stability, University of Chicago Press 1966.
12 CHANDRASEKHAR, S., Hydrodynamic and Hydromagnetic Stability, Clarendon Prcss, Oxford 1961.
13 GOUGH, D. O., SPIEGEL, E. A., and TOOMRE, J., To appear.
14 MURPHY, J. O., Proc. Astron. Soc. Aust. **2**, p. 53 (1971).
15 STEINER, J. M., Zastosowania Matematyki. **41**, p. 271 (1973).
16 BHATIA, P. K. and STEINER, J. M., ZAMM **52**, S. 321 (1972).
17 BHATIA, P. K. and STEINER, J. M., Phys. Lett. A **37**, p. 419 (1971).
18 BHATIA, P. K. and STEINER, J. M., J. Math. Anal. and Appl. **41**, p. 271 (1973)
19 BHATIA, P. K. and STEINER, J. M., Aust. J. Phys. **25**, p. 695 (1972).
20 FOSTER, T. D., J. Fluid Mech., **37**, p. 81 (1969).
21 FOSTER, T. D., Geoph. Fluid Dynamics **2**, p. 201 (1971).
22 HERRING, J. R., J. Atmos. Sci. **21**, p. 277 (1964).

Eingereicht am 13. 10. 1972, revid. Fassung 6. 4. 1973

Anschrift: R. VAN DER BORGHT, J. O. MURPHY and J. M. STEINER, Department of Mathematics, Monash University, Melbourne, Australia

ZAMM 54, 9–17 (1974)

Dj. S. Djukić*)

The Universal Form of the Boundary Layer Equation and its Solution for Non-Newtonian Power-Law Fluids

Die vorliegende Arbeit bringt eine theoretische Untersuchung der laminaren Grenzschichtströmung einer dem Potenzgesetz gehorchenden nichtnewtonschen Flüssigkeit hinter einer beliebigen zweidimensionalen Fläche. Es wird eine Form der Grenzschichtgleichung hergeleitet, die in der Hinsicht universell ist, daß weder die Gleichung noch die Randbedingungen von den besonderen Bedingungen des Problems abhängen. Dies wird dadurch erreicht, daß in die Zahl der unabhängigen Variablen eine Schar von Parametern einbezogen wird, die den Einfluß der für ein spezielles Problem charakteristischen Bedingungen ausdrücken. Die universelle Gleichung wird durch eine Parameterapproximation numerisch gelöst. Mit Hilfe dieser Lösung werden mehrere spezielle Probleme im Detail diskutiert.

This paper presents a theoretical analysis of the laminar boundary layer flow of a non-Newtonian fluid, obeying the power-law model, past an arbitrary two-dimensional surface. The universal form of the boundary layer equation, such that neither the equation nor the boundary conditions depend on the particular conditions of the problem, is derived. This universality is achieved by introducing a set of parameters expressing the influence of external conditions characteristic of a particular problem into a number of independent variables. The universal equation is solved numerically by a parameter approximation. Using this solution several particular problems are analysed in detail.

В настоящей работе приведено теоретическое исследование ламинарного течения пограничного слоя неньютоновской жидкости, подчиняющейся степенному закону, за любой двухмерной поверхностью. Форма выведенного уравнения пограничного слоя является универсальной в том смысле, что ни само уравнение ни граничные условия не зависят от особых условий задачи. Это достигается путём введения в число независимых переменных семейства параметров, учитывающих влияние характеристических условий для каждой специальной задачи. Универсальное уравнение решается численно путём аппроксимации параметров. С помощью этого решения обсуждаются подробно многие специальные задачи.

1. Introduction

Considerable attention has recently been devoted to predicting the behaviour of non-Newtonian fluids in motion. This is due to the fact that fluids, such as molten plastics, pulps, slurries, emulsions, etc., are being produced industrially in increasing quantities, and do not obey the Newtonian postulate that the stress tensor is directly proportional to the deformation tensor.

It is well known, though, that a chief difficulty in the study of non-Newtonian fluid mechanical phenomena is that to this point no definite relationship between the stress tensor and the deformation tensor, valid for all fluids, has been discovered. Thus, except for simple cases, a generalized form of the Navier-Stokes equations, obeyed by all fluids in motion, can not be formulated, and so theoretical studies in this area have been limited not so much by the mathematical complexity of the basic non-linear equations as by the inability to arrive at their correct form.

We shall restrict our analysis here to a special family of non-Newtonian substances, the so-called power-law fluids, with the two-parameter non-linear equation of state

$$\tau_{ij} = -p\,\delta_{ij} + k\,|\dot e_{ms}\,\dot e_{sm}/2|^{\frac{n-1}{2}}\,\dot e_{ij}\,, \tag{1.1}$$

which is adequate for many non-Newtonian fluids. Here, τ_{ij} — stress tensor, p — pressure, δ_{ij} — Kronecker delta, k — coefficient of consistency, $\dot e_{ms}$ — strain rate tensor and n — power-law flow behaviour index. When $n > 1$, the fluid is dilatant; $n < 1$ denotes a pseudoplastic fluid. In this paper it will be assumed that k and n are constants.

The flow of non-Newtonian fluids in pipes and channels has been studied primarily because of the importance and relative simplicity of the phenomena (Metzner and Reed [30], Frederickson and Bird [13], Wilkinson [42], Kapur and Gupta [20], Gupta [15]).

The boundary layer concept in the theory of non-Newtonian power-law fluids was introduced by Schowalter [35] and similar solutions were analysed by him, Kapur and Srivastava [21], Lee and Ames [24] and most successfully by Schullman and Berkovsky [40]. The flow across a flat plate was studied by Acrivos, Shah and Petersen [3]–[4], Schullman and Berkovsky [40], Lemieux, Dubey and Unny [26]. The two-dimensional jet flow problem was solved by Kapur [19] and Gutfinger and Shinnar [16]. The two-dimensional stagnation point flow problem was analysed by Schullman and Berkovsky [40]. The solution to Stokes' first problem was obtained by Lemieux, Unny and Dubey [27] and Rott [32]. Cebeci, Clauss and Dyer [8] solved the problem of Howarth flow [18] for this type of fluid.

From the papers cited, on the boundary layer flow of non-Newtonian power-law fluids, it follows that there exist only a small number of solutions to particular problems in this area.

In solving the boundary layer problems for a non-Newtonian power-law fluid Djordjevic [10] and Serth and Kiser [36] adapted the Görtler [14] series method. While Djordjevic obtained a formal solution the former authors solved for

*) Mechanical Engineering Faculty, University of Novi Sad, 21000 Novi Sad, Yugoslavia

a number of universal functions, but this method did not allow them to reach the separation point. DJORDJEVIC [10], also formally, developed the method of LOITSANSKII [28]. Approximative methods, for boundary layers of the power-law fluids, were constructed by ACRIVOS, SHAH and PETERSEN [3]—[4], BIZZEL and SLATTERY [6], LEE and AMES [24], SCHULLMAN and BERKOVSKY [40] and DORFMAN and VISNEVSKII [11]—[12].

The purpose of this work is to provide analytical solutions and qualitative discussions of the laminar boundary layer flow of non-NEWTONian fluids, which obey a power-law model, past an arbitrary two-dimensional surface. To this end the method of LOITSANSKII [28] for the NEWTONian case, developed by KAPUSTYANSKII [22], BOGDANOVA [7], CHAN [9], DJORDJEVIC [10], KARYAKIN [23], LYUBENOV [29] and generalized by SALNIKOV [33], will be used.

2. Governing Equations and Boundary Conditions

For the problem of plane steady laminar boundary layer flow of power-law fluid SCHOWALTER [35] derived the following system of differential equations

$$u\,u_x + v\,u_y = V\,V_x + \frac{k}{\varrho}\,[|u_y|^{n-1}\,u_y]_y \tag{2.1}$$

$$u_x + v_y = 0\,, \tag{2.2}$$

with the corresponding boundary conditions

$$\begin{aligned} u = v &= 0 && \text{for} \quad y = 0\,, \\ u &= V(x) && \text{for} \quad y = \infty\,, \\ u &= u_0(y) && \text{for} \quad x = x_0\,, \end{aligned} \tag{2.3}$$

where: x — distance along the surface from the leading edge, y — distance normal to the surface, u — velocity component along x, v — velocity component along y, ϱ — density, $V(x)$ — velocity of external flow in a direction parallel to the wall at the edge of the boundary layer (in this paper it will be assumed that this velocity is equal to the velocity of non-viscous flow around the same body), $u_0(y)$ — the known profile of velocity u in some given section $x = x_0$ of the boundary layer, x, y — as a lower index indicates partial differention with x and y respectively. These differential equations are valid for sufficiently large values of the generalized REYNOLDS number for the flow of power-law fluids.

In general, the velocity component u will increase from its zero value at the wall to the value $V(x)$ at the edge of the boundary layer and thus in this case u_y will be non-negative ($u_y \geqq 0$). For this reason the absolute sign appearing in the viscous term of equation (2.1) will be dropped in further considerations.

Introducing the stream function Ψ, defined by

$$u = \Psi_y\,, \qquad v = -\,\Psi_x\,, \tag{2.4}$$

the continuity equation (2.2) is identically satisfied while the system (2.1) and (2.3) transforms into

$$\Psi_y\,\Psi_{xy} - \Psi_x\,\Psi_{yy} = V\,V_x + \nu[(\Psi_{yy})^n]_y\,, \tag{2.5}$$

$$\begin{aligned} \Psi_y = \Psi_x &= 0 && \text{for} \quad y = 0\,, \\ \Psi_y &= V(x) && \text{for} \quad y = \infty\,, \\ \Psi_y &= u_0(y) && \text{for} \quad x = x_0\,, \end{aligned} \tag{2.6}$$

where $\nu = k/\varrho$ is constant.

3. Transformation of the Boundary Layer Equation

For the analysis of steady flow of a power-law fluid in the boundary layer, let us introduce the new independent variables x and η instead of the old ones x and y

$$x = x\,, \qquad \eta = \frac{B(x)}{\delta^{**}(x)}\,y\,, \tag{3.1}$$

and a dimensionless stream function F

$$\Psi(x, y) = \frac{\delta^{**}(x)}{B(x)}\,V(x)\,F(x, \eta)\,, \tag{3.2}$$

where δ^{**} is the momentum thickness

$$\delta^{**}(x) = \int\limits_0^\infty \frac{u}{V}\left(1 - \frac{u}{V}\right)dy\,, \tag{3.3}$$

and, by substitution of (3.1—2) into (3.3), we see that

$$B(x) = \int_0^\infty F_\eta (1 - F_\eta)\, d\eta\,, \qquad \left(F_\eta \equiv \frac{dF}{d\eta}\right). \tag{3.4}$$

By means of (3.1—2) the differential equation (2.5) and the corresponding boundary conditions (2.6) become

$$[(F_{\eta\eta})^n]_\eta + \frac{f_1}{B^{n+1}}(1 - F_\eta^2) + \left[\frac{V}{n+1}\left(\frac{f_1}{B^{n+1} V_x}\right)_x + \frac{2\, n f_1}{(n+1)\, B^{n+1}}\right] F\, F_{\eta\eta}$$

$$= \frac{V f_1}{B^{n+1} V_x}(F_\eta F_{\eta x} - F_x F_{\eta\eta})\,, \tag{3.5}$$

$$F_\eta = 0\,, \qquad \left(\frac{\delta^{**} V}{B}\right)_x F + \frac{\delta^{**} V}{B} F_x = 0\,, \qquad \eta = 0\,,$$

$$F_\eta = 1\,, \qquad \eta = \infty\,, \tag{3.6}$$

$$F_\eta = F_{0\eta}(\eta)\,, \qquad x = x_0\,.$$

Here

$$f_1(x) = \frac{(\delta^{**})^{n+1} V_x}{\nu\, V^{n-1}}\,, \qquad (\nu = k/\varrho)\,, \tag{3.7}$$

is the form-parameter (see SCHULLMAN and BERKOVSKII [40] p. 130) for the boundary layer analysis of a power-law fluid. This parameter characterises the development of the velocity field in the boundary layer along the x-direction.

4. On the Momentum Equation for the Boundary Layer

The momentum equation for the boundary layer of a power-law fluid has the form [40] p. 131

$$\frac{df_1}{dx} = \frac{V_x}{V}\Phi + \frac{V_{xx}}{V_x} f_1 \quad \text{or} \quad \frac{dz^{**}}{dx} = \frac{\Phi}{V}\,, \qquad (z^{**} = f_1/V_x)\,, \tag{4.1}$$

where the form-parameter f_1 is given by (3.7) and the well known boundary layer quantities are introduced (see for example [40] p. 131 or [43] p. 627)

$$\Phi = (n+1)\left[\zeta - \left(2 + H + \frac{n-1}{n+1}\right) f_1\right], \tag{4.2}$$

$$\zeta = \left\{\left[\frac{\partial(u/V)}{\partial(y/\delta^{**})}\right]^n\right\}_{y=0}; \qquad H = \delta^*/\delta^{**}; \qquad \delta^* = \int_0^\infty \left(1 - \frac{u}{V}\right) dy\,. \tag{4.3}$$

Combining (2.4), (3.1—2) and (3.7) with the previous expressions we have

$$\zeta = B^n[(F_{\eta\eta})^n]_{\eta=0}\,; \qquad H = A/B\,; \qquad \delta^* = \delta^{**} A/B\,, \tag{4.4}$$

where

$$A(x) = \int_0^\infty (1 - F_\eta)\, d\eta\,. \tag{4.5}$$

In the following analysis, let us introduce the following [43] p. 634 infinite set of form-parameters

$$f_k = V^{k-1} \frac{d^k V}{dx^k} (z^{**})^k\,, \qquad (k = 1, 2, \ldots)\,, \tag{4.6}$$

where the first term ($k = 1$; $f_1 = V_x z^{**}$) is defined above. Differentiating (4.6) with respect to x we see that the relation

$$\frac{V}{V_x} f_1 \frac{df_k}{dx} = [(k-1) f_1 + k\, \Phi] f_k + f_{k+1} = \theta_k\,, \qquad (k = 1, 2, \ldots) \tag{4.7}$$

is valid.

The function $V(x)$ is an arbitrary function, so the parameters f_k [43] p. 634, forms a set of independent functions. With z^{**} all form-parameters depend on the boundary layer thickness. The first parameter f_1 characterizes the inclination of the line $V(x)$. For positive values of f_1 the flow is accelerated and for negative f_1 decelerated. The second form-parameter f_2 takes into account the curvature of the function $V(x)$. The form-parameters of higher order do not have such obvious geometric interpretations, but their influence is larger if higher order derivatives of the function $V(x)$ have larger values in the interval of the variable x under consideration.

5. The Universal Equation for the Boundary Layer of a Power-Law Fluid

In the general case the solution of the differential equation (3.5) and the corresponding boundary conditions (3.6) is a function of the variables x, η, the function $V(x)$ and the non-NEWTONian flow behaviour index n, i.e. $F = F(x, \eta, \{V(x)\}, n)$. Introducing the set of independent variables (4.6) instead of the independent variable x we have

$$F(x, \eta, \{V(x)\}, n) = F(\eta, n, f_1, f_2, \ldots) . \tag{5.1}$$

From (4.7) it follows that the connection between the partial derivative with respect to x and with respect to the set of the variables f_k is

$$\frac{\partial}{\partial x} = \frac{V_x}{V f_1} \sum_{k=1}^{\infty} \theta_k \frac{\partial}{\partial f_k} . \tag{5.2}$$

Now, using (4.1), (3.7) and (5.1–2) the differential equation (3.5) and boundary conditions (3.6) assume the final form

$$[(F_{\eta\eta})^n]_\eta + \frac{\Phi(f_k) + 2nf_1}{(n+1)B^{n+1}} F F_{\eta\eta} + \frac{f_1}{B^{n+1}}(1 - F_\eta^2) = \frac{1}{B^{n+1}} \sum_{k=1}^{\infty} \theta_k \left(F_\eta F_{\eta f_k} - F_{\eta\eta} F_{f_k} + \frac{1}{B} B_{f_k} F F_{\eta\eta}\right). \tag{5.3}$$

$$\begin{aligned} &F_\eta = 0 , \qquad \frac{\Phi + 2nf_1}{n+1} F + \sum_{k=1}^{\infty} \theta_k \left(F_{f_k} - \frac{F}{B} B_{f_k}\right), \qquad \eta = 0; \\ &F_\eta = 1 , \qquad \eta = \infty; \\ &F_\eta = F_{0\eta}(\eta) , \qquad f_k = f_{k0} . \end{aligned} \tag{5.4}$$

In the non-linear partial differential equation (5.3) and in the corresponding boundary conditions (5.4) the velocity of the external flow on the border of the boundary layer does not appear, and for this reason this equation is the universal differential equation for the plane steady laminar boundary layer flow of a non-NEWTONian power-law fluid. For NEWTONian fluids ($n = 1$) and eq. (5.3) reduces to the universal eq. obtained by SALNIKOV [33]. If we make the assumption that the function B, given by (3.4), is constant and the fluid NEWTONian, then eq. (5.3) transforms into the universal eq. obtained by LOITSANSKII [28].

Here, we must remember that the function $F_0(\eta)$ in the boundary conditions (5.4) is a known velocity profile for u in the section $f_k = f_{k0}$ of the boundary layer. In the space of variables $f_1, f_2, \ldots$, the point $f_1 = f_2 = \cdots = 0$ is an singular point of the eq. (5.3). At this point we can require that the function $F_0(\eta)$ is equal to the solution of the problem of the flow of a power-law fluid across a flat plate, defined by (ACRIVOS, SHAH and PETERSEN [3]–[4])

$$\begin{aligned} &n(F_{0\eta\eta})^{n-1} F_{0\eta\eta\eta} + F_0 F_{0\eta\eta} = 0 , \\ &F_0(0) = 0 , \qquad F_{0\eta}(0) = 0 , \qquad F_{0\eta}(\infty) = 1 . \end{aligned} \tag{5.5}$$

For $f_1 = f_2 = \cdots = 0$, from (4.7) it follows that $\theta_1 = \theta_2 = \cdots = 0$. Also, in this case the function $B(x)$, given by (3.4), is constant $(F_\eta = F_{0\eta}(\eta))$ and will be denoted by B_0. Hence, the solution of the system (5.3–4), when $f_1 = \cdots = 0$, will be same as that obtained from (5.5) only if the following condition is satisfied

$$B_0 = [(F_{0\eta\eta})^n]_{\eta=0} , \tag{5.6}$$

where (4.2) and (4.4) was used.

6. Numerical Solution of the Universal Equation

The boundary layer flow of a non-NEWTONian power-law fluid problem can be solved by applying the universal eq. (5.3) with the boundary conditions (5.4). This differential equation is a non-linear partial differential equation of the third order whose solution, the function F, is a function of the variable η and the infinite set of variables $f_1, f_2, \ldots$ Imposing the condition that the first k variables of the set f_k are different from zero and all others equal to zero we have, from (5.3), the k-th parameter approximation of the full universal equation (for more details see [28]).

Further, let us consider the one-parameter approximation of the universal equation (5.3). Let all parameters f_k, except the first one, be equal to zero, i.e. $f_1 \neq 0, f_2 = f_3 = \cdots = 0$. Now, from (4.7) it follows that: $\theta_1 = f_1 \Phi$, $\theta_2 = \theta_3 = \cdots = 0$. Combining these results with (5.3–4) we obtain the single-parameter approximation of the universal eq. and the corresponding boundary conditions

$$\begin{aligned} &[(F^{(1)}_{\eta\eta})^n]_\eta + \frac{\Phi^{(1)} + 2nf_1}{(n+1)(B^{(1)})^{n+1}} F^{(1)} F^{(1)}_{\eta\eta} + \frac{f_1}{(B^{(1)})^{n+1}} [1 - (F^{(1)}_\eta)^2] \\ &\quad = \frac{f_1 \Phi^{(1)}}{(B^{(1)})^{n+1}} \left[F^{(1)}_\eta F^{(1)}_{\eta f_1} - F^{(1)}_{\eta\eta} F^{(1)}_{f_1} + \frac{F^{(1)}}{B^{(1)}} B^{(1)}_{f_1} F^{(1)}_{\eta\eta}\right] , \end{aligned} \tag{6.1}$$

$$\begin{aligned} &F^{(1)}_\eta = 0 , \qquad \frac{\Phi^{(1)} + 2nf_1}{n+1} F^{(1)} + f_1 \Phi^{(1)} \left[F^{(1)}_{f_1} - \frac{F^{(1)}}{B^{(1)}} B^{(1)}_{f_1}\right] = 0 , \qquad \eta = 0; \\ &F^{(1)}_\eta = 1 , \quad \eta = \infty; \qquad F^{(1)}_\eta = F_{0\eta}(\eta) , \quad f_1 = 0 , \end{aligned} \tag{6.2}$$

where $F_0(\eta)$ is solution of the system (5.5). Here, and in the coming analyses the upper index (1) denotes the quantities calculated from the single-parameter approximation of the universal equation.

The differential equation (6.1) with the corresponding boundary conditions (6.2) ought to be solved along with the equations (4.2), (4.4–5) and (3.4). It is obvious that an analytical solution of this problem can not be found. The numerical approach to this difficult mathematical problem makes it evident that the solution is unstable. The solution becomes stable if an approximation to the complete problem is made, namely that the function $B^{(1)}$, which depends on f_1, is constant. We suppose that this function is equal to the same function-constant calculated for the power-law fluid flow across a flat plate

$$B^{(1)}(f_1) = B_0\,, \tag{6.3}$$

where B_0 is given by (5.6). This assumption is equivalent to that made by LOITSANSKII [28] in considering NEWTONian boundary layers. The necessary values of $F_{0\eta\eta}(0)$, for calculating the constant B_0 by (5.6), are taken from ACRIVOS, SHAH and PETERSEN [3] for use below.

The problem of the boundary layer flow now consists of solving equations (6.1–2), (4.2), (4.4–5), (3.4), (5.6) and (6.3). The numerical integration method for non-linear partial differential equations (see for instance LOITSANSKII [28], KARYAKIN [23], LYUBENOV [29]) was adapted for this system of equations. The stable numerical solution of these eq. was obtained on an IBM-360 digital computer for the following values of the power-law flow behaviour index: $n = 0{,}5$; $1{,}5$; 2; $2{,}5$; 3; 4 and 5. The solution for the NEWTONian case $(n = 1)$ is same as that obtained by LOITSANSKII [28]. For $n < 0{,}5$ a stable solution in the separation point vicinity could not be found. The numerical integration of the equations was performed in the interval $f_{1s} \leqq f_1 \leqq f_{1m}$, where f_{1s} is the value of the formparameter at the separation point, i.e. $\zeta^{(1)}(n, f_{1s}) = 0$, and f_{1m} is the value of f_1 at the front stagnation point of the fluid flow, i.e. $\Phi^{(1)}(n, f_{1m}) = 0$. Calculated characteristic functions for the boundary layer analysis $\zeta^{(1)}(n, f_1)$, $H^{(1)}(n, f_1)$ and $\Phi^{(1)}(n, f_1)$ are plotted in the figures 1–4.

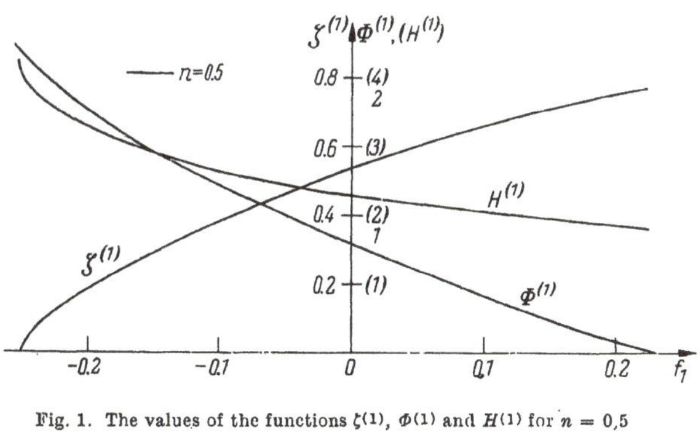

Fig. 1. The values of the functions $\zeta^{(1)}$, $\Phi^{(1)}$ and $H^{(1)}$ for $n = 0{,}5$

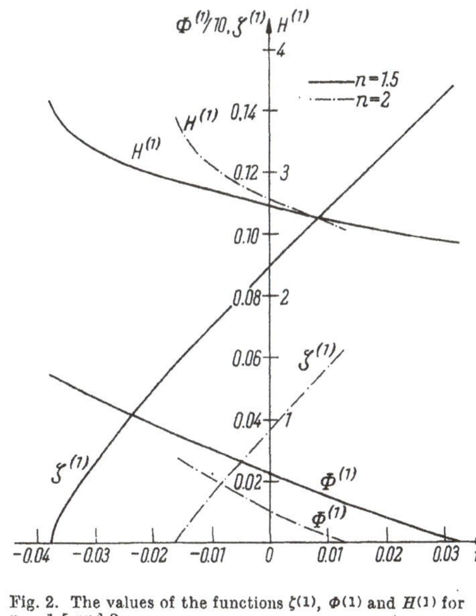

Fig. 2. The values of the functions $\zeta^{(1)}$, $\Phi^{(1)}$ and $H^{(1)}$ for $n = 1{,}5$ and 2

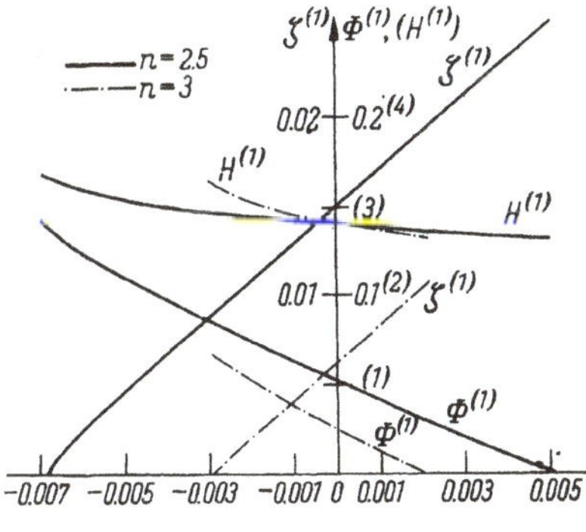

Fig. 3. The values of the functions $\zeta^{(1)}$, $\Phi^{(1)}$ and $H^{(1)}$ for $n = 2{,}5$ and 3

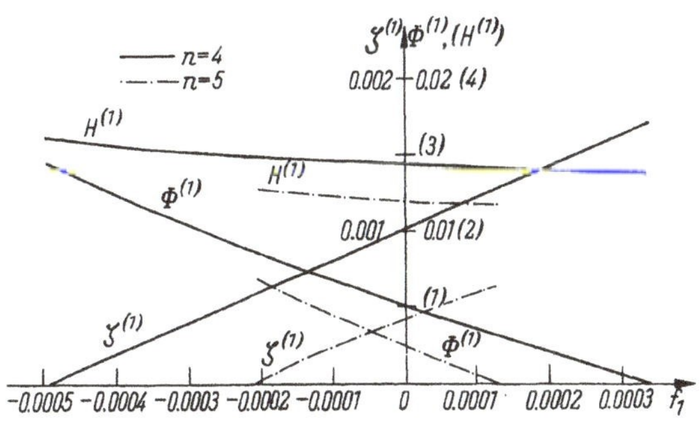

Fig. 4. The values of the functions $\zeta^{(1)}$, $\Phi^{(1)}$ and $H^{(1)}$ for $n = 4$ and 5

The values for f_{1s}, f_{1m} and the values Φ_0, Φ_1 and Φ_2 of the power series

$$\Phi^{(1)}(n, f_1) = \Phi_0 - \Phi_1 f_1 + \Phi_2 (f_1)^2 - \cdots, \tag{6.4}$$

which will be used below, are given in table 1 for different values of the parameter n.

Table 1. The values of f_{1s}, f_{1m}, Φ_0, Φ_1 and Φ_2 as functions of the parameter n

n	0,5	1	1,5	2	2,5	3	4	5
Φ_0	0,802185	0,4408	0,22497	0,11007	0,052537	0,024737	$0{,}50982 \cdot 10^{-2}$	$0{,}25063 \cdot 10^{-2}$
Φ_1	+4,467	+5,7139	+7,638	+9,524	+11,344	+13,02	+16,675	+20,14
Φ_2	3,931	6,0189	24,32	72,001	204,1	632,41	4374,4	6475
f_{1s}	−0,2512	−0,0852	−0,0376	−0,0162	−0,006835	−0,002925	−0,000495	$-0{,}2068 \cdot 10^{-3}$
f_{1m}	0,22391	0,0854	0,03286	0,01284	0,005105	0,00212	0,000335	$0{,}13 \cdot 10^{-3}$

7. Integration of the Momentum Equation

The numerical solution of the universal equation can be applied to specific boundary layer problems if we have the functional dependence of the form-parameter f_1 on the variable x. Thus, we must integrate the momentum equation (4.1), which is a first order non-linear differential equation. If we neglect the terms $(f_1)^2, (f_1)^3, \ldots$ in equation (6.4) and then the function $\Phi^{(1)}$ is substituted into (4.1) and integrated, we get

$$f_1(x) = \frac{\Phi_0 \, V_x(x) \int_0^x V(t)^{\Phi_1 - 1} \, dt}{V(x)^{\Phi_1}}, \tag{7.1}$$

where the boundary conditions are: $f_1 = f_{1m}$ for $x = 0$, in the case of the bodies with a front stagnation point, and $f_1 = 0$ for $x = 0$ when the body has a sharp front edge. The function $\Phi^{(1)}$ deviates from the linear law only in a small area near the separation point (see fig. 1—4). This deviation is neglible and so, the linearisation is completely justified in the entire interval of the variable f_1.

8. Examples

The solutions of the universal differential equation and the momentum equation, can be used to solve special plane steady laminar boundary layer problems involving non-Newtonian power-law fluids. The problems can be carried out in the interval $0{,}5 \leqq n \leqq 5$ of the power-law fluid index.

The aim of the following examples is the study of boundary layer magnitudes, such as the separation point position, the boundary layer thickness and the shearing stress. So far as possible the comparison of the results with other methods will be given.

Further, all magnitudes with index 1 are dimensionless, and the lower index s indicates the value at the separation point.

8.1 The separation point position

Using equation (7.1) and table 1, the separation point position x_{1s} of the boundary layer is calculated for different functions $V_1(x_1)$ and for various values of the non-Newtonian behaviour parameter n and plotted in fig. 5. Also, in fig. 5 the values p_s of the parameter p, for the external velocity distribution $V_1 = (x_1)^p$, is given, when the shearing stress is zero at all points of the body. In fig. 5 the results for $V_1 = 1 - x_1$ are compared to two other solutions of the same problem, the numerical solution [8] and the approximate solution [3]. The results of this method are in good agreement with the numerical solution.

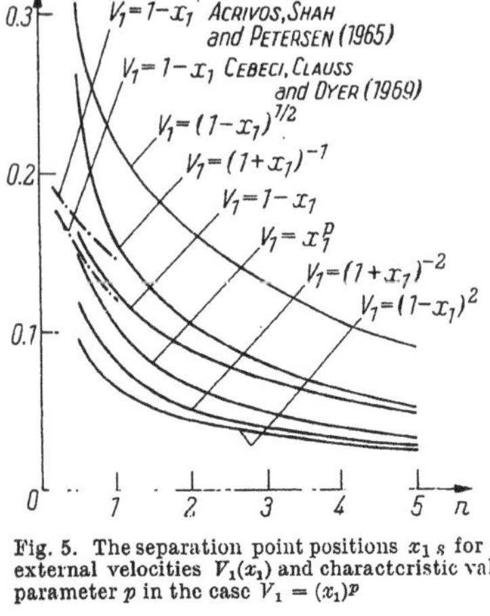

Fig. 5. The separation point positions x_{1s} for different external velocities $V_1(x_1)$ and characteristic values p_s of the parameter p in the case $V_1 = (x_1)^p$

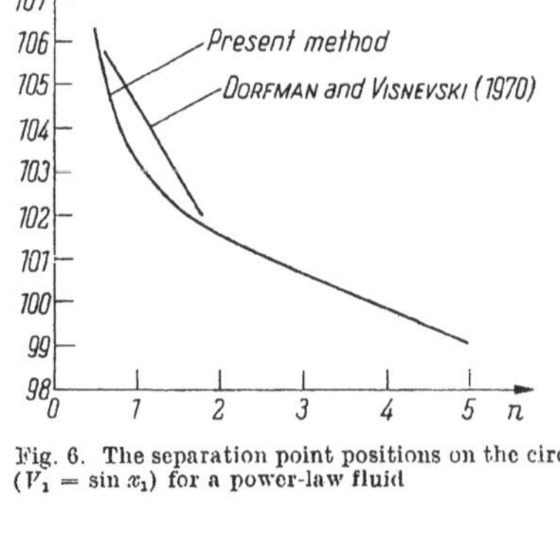

Fig. 6. The separation point positions on the circular cylinder ($V_1 = \sin x_1$) for a power-law fluid

In fig. 6 the separation point positions are given for power-law fluid flow around a circular cylinder and compared with the approximate solution of DORFMAN and VISNEVSKII [12].

From figures 5 and 6 it is possible to conclude that increasing the parameter n causes separation point of these boundary layers to move upstream.

For further comparison the present method's results, for x_{1s} and p_s, are compared to the results of other authors, in the case of a NEWTONian fluid ($n = 1$):

$V_1 = (x_1)^p$, $p_s = -0{,}1012$, $p_s = -0{,}091$ [17]; $V_1 = 1 - x_1$, $x_{1s} = 0{,}123$, $x_{1s} = 0{,}1198$ [25];
$V_1 = (1 + x_1)^{-1}$, $x_{1s} = 0{,}157$, $x_{1s} = 0{,}159$ [14]; $V_1 = (1 + x_1)^{-2}$, $x_{1s} = 0{,}089$, $x_{1s} = 0{,}078$ [14];
$V_1 = (1 - x_1)^{1/2}$, $x_{1s} = 0{,}22$, $x_{1s} = 0{,}223$ [14]; $V_1 = (1 - x_1)^2$, $x_{1s} = 0{,}065$, $x_{1s} = 0{,}067$ [14];
$V_1 = \sin x_1$, $x_{1s} = 103{,}1°$, $x_{1s} = 104{,}5°$ [41].

8.2. The boundary layer thickness

Consider the flow of a power-law fluid with external velocity [34] p. 163

$$V = V_\infty - b\,x\,, \tag{8.1}$$

where: V_∞ — constant velocity, b — negative or positive constant. Using (3.7), (7.1) and (8.1) we have the dimensionless momentum thickness

$$\delta_1^{**} = R_e^{-\frac{1}{n+1}}\left\{\frac{\Phi_0\,[1-(1-x_1)^{\Phi_1}]}{\Phi_1\,(1-x_1)^{\Phi_1+1-n}}\right\}^{\frac{1}{n+1}}, \tag{8.2}$$

where $x_1 = x/(V_\infty/b)$, $\delta_1^{**} = \delta^{**}/(V_\infty/b)$ and $R_e = (V_\infty)^{2-n}\,(V_\infty/b)^n/\nu$ is the generalized REYNOLDS number for a power-law fluid. In fig. 7 the momentum thickness (8.2) for various values of the parameter n and for $R_e = 300$, is given.

For the same values of the generalized REYNOLDS number and same values of coordinate x_1, in this figure, it is obvious that the dimensionless momentum thickness grows with the increment of the non-NEWTONian parameter n.

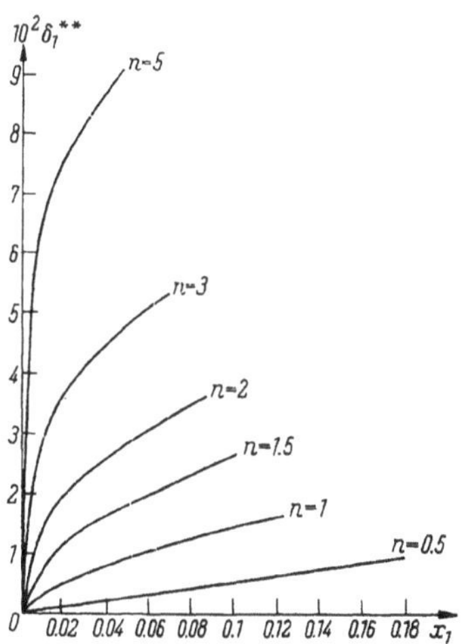

Fig. 7. The dimensionless momentum thickness for the external velocity $V_1 = 1 - x_1$ and $R_e = 300$

For the flow around a circular cylinder the external velocity is

$$V(x) = 2\,V_\infty \sin(x/R)\,, \tag{8.3}$$

where: R — radius of cylinder, V_∞ — free stream velocity. Combining (8.3) and (3.7), we have the dimensionless momentum thickness

$$\delta_1^{**} = \left[\frac{f_1(x_1)\,(\sin x_1)^{n-1}}{\cos x_1}\right]^{\frac{1}{n+1}}, \tag{8.4}$$

where

$$x_1 = x/R\,, \qquad \delta_1^{**} = \delta^{**}(R_e)^{\frac{1}{n+1}}/R \tag{8.5}$$

and the generalized REYNOLDS number is

$$R_e = (2\,V_\infty)^{2-n}\,R^n/\nu\,. \tag{8.6}$$

When $x_1 \to 0$, from (7.1) and (8.3), it follows that $f_1 \to \Phi_0/\Phi_1$, and then, from (8.4), $\delta_1^{**} \to 0$ for $n > 1$ (dilatant fluid) and $\delta_1^{**} \to \infty$ for $n < 1$ (pseudoplastic fluid). The equation of state (1.1) for a power-law fluid can be written as $\tau_{ij} = -p\,\delta_{ij} + \mu_e\,\dot{e}_{ij}$, where μ_e is the effective viscosity

$$\mu_e = k\,|\dot{e}_{mp}\,\dot{e}_{pm}/2|^{\frac{n-1}{2}}\,.$$

Near the front stagnation point on the circular cylinder $\dot{e}_{ij} \approx 0$ and then $\mu_e \to 0$ for $n > 1$ and $\mu_e \to \infty$ for $n < 1$. From this analysis we see that for $n > 1$, at the front stagnation point on the circular cylinder, the effective viscosity and momentum thickness tend to zero while for $n < 1$ they tend to infinity. It is certain that in real non-NEWTONian power-law fluid flow such extremal values of the boundary layer thickness can not be achieved. Hence, the conclusion is that the power-law equation of state yields contradictory results in flow districts where the strain rate tensor has small values. This confirms the suppositions of REINER [31].

8.3 The shearing stress

For flow around a circular cylinder, combining (2.4), (3.1–2), (8.3–6) we have the dimensionless magnitude

$$\left[\left(\frac{\partial u_1}{\partial y_1}\right)^n\right]_{y_1=0} = \left(\frac{V_1}{\delta_1^{**}}\right)^n \zeta\,, \tag{8.7}$$

where ζ is given by (4.4), $V_1 = V/2V_\infty$, $u_1 = u/2V_\infty$ and $y_1 = y(R_e)^{\frac{1}{n+1}}/R$. Now, from (8.5) and the expression for the shearing stress on the body $\tau_0 = k\,[(\partial u/\partial y)^n]_{y=0}$, we obtain the friction coefficient on the circular cylinder

$$C_f = R_e^{-\frac{1}{n+1}}\left[\left(\frac{\partial u_1}{\partial y_1}\right)^n\right]_{y_1=0}, \qquad (c_f = \tau_0/4\,\varrho\,V_\infty)\,.$$

In fig. 8 and 9 the dimensionless magnitude (8.7) and the friction coefficient for generalized REYNOLDS number $R_e = 600$ are given.

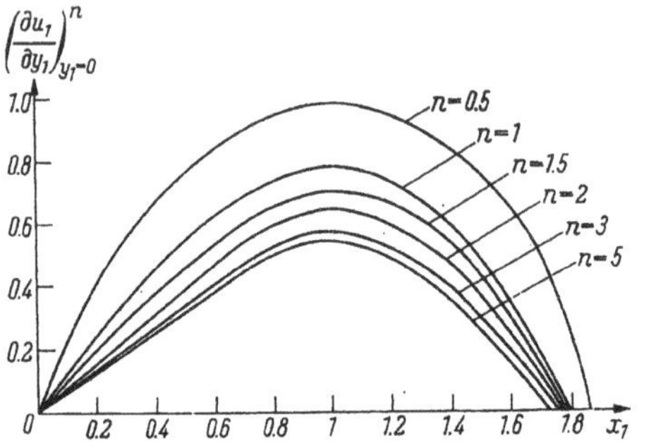

Fig. 8. The dimensionless magnitude $[(\partial u_1/\partial y_1)^n]_{y_1=0}$ for power-law fluid flow around a circular cylinder

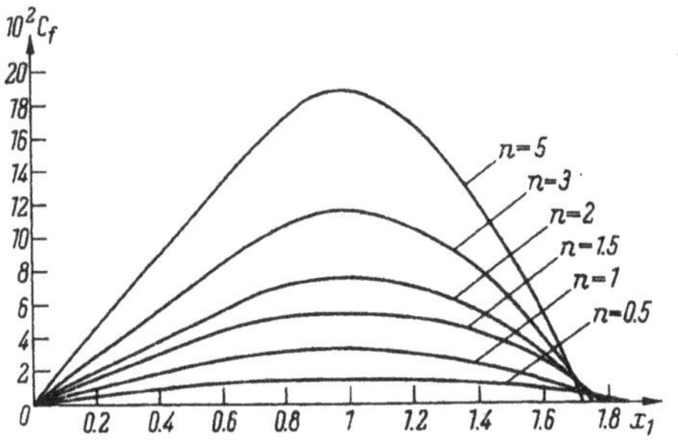

Fig. 9. The friction coefficient for $R_e = 600$ on the circular cylinder in the power-law stream

From the fig. 8 it is obvious that the friction coefficient on the circular cylinder increases with increasing values of the power-law fluid parameter n.

9. Conclusions — The Method

The results obtained by means of the universal equation of the boundary layer for a power-law fluid, are in good agreement with the results of other authors.

According to LOITSANSKII [28] some picture of the exactness of the solution to the universal equation in the one-parameter approximation can be obtained by an analysis of higher order form-parameters in the interval $0 \leqq x \leqq x_s$. Namely, if the following conditions are satisfied for every $0 \leqq x \leqq x_s$

$$|f_2|_{\max} \ll \varepsilon\,, \qquad |f_3|_{\max} \ll \varepsilon\,,$$

where $\varepsilon = |f_1|_{\max}$, then neglecting the form-parameters $f_2, f_3, \ldots$, in the universal equation is completely justified, and the one-parameter approximation is very close to the exact solution of the same problem.

For example, the analysis in the case of power-law fluid flow around a circular cylinder shows that the one-parameter solution, for $n > 1$ (dilatant fluid) is very close to the exact one, whereas for $n < 1$ (pseudoplastic fluid) there is greater error.

The exactness of the present method can be improved by a higher order parameter approximation of the universal equation and by solution of the momentum equation in the non-linear form.

10. Conclusion — The Physical Behaviour of Boundary Layer Flow for a Power-Law Fluid

From the problems discussed one can come to some conclusions about the physical properties of non-NEWTONian power-law boundary layer flow.

a) The separation point coordinate x_s of the boundary layer decreases with the increase of the non-NEWTONian power-law fluid parameter n.

b) Fixing the value of the generalized REYNOLDS number and increasing the parameter n causes the boundary layer on the same body to thicken.

c) Increasing the parameter n causes the friction coefficient on the body, which moves through the power-law fluid, to increase.

d) These conclusions are completely in the spirit of the well known general behaviour of boundary layers. Namely, that the increase of the boundary layer thickness causes larger friction resistance and smaller separation point coordinate.

Acknowledgements

The author wishes to express his deep sense of gratitude to Professor V. SALNIKOV for constant help and guidance throughout this investigation. I am also very grateful to A. STRAUSS for his constructive review of the manuscript and to the referee for his very useful suggestions. The numerical calculations were performed on the IBM-360 Model 44 digital computer of the Mathematical Institute of Belgrade.

References

1 Acrivos, A., A Rapid Method for Estimating the Shear Stress and the Separation Point in Laminar Incompressible Boundary-Layer Flows, Jour. of the Aero-Space Sci. April, 314–315 (1960).
2 Acrivos, A., Inst. Chem. Eng. J. 6, 584 (1960).
3 Acrivos, A., Shah, M. J., Petersen, E. E., On the Solution of the Two-dimensional Boundary–Layer Flows Equations for a Non-Newtonian Power-Law Fluid, Chem. Eng. Sci. **20**, 101–105 (1965).
4 Acrivos, A., Shah, M. J., Petersen, E. E., Momentum and Heat Transfer in Laminar Boundary-Layer Flow of Non-Newtonian Fluids Past External Surfaces, A. I. Ch. E. Jour. **6**, 312–317 (1960).
5 Bird, R. B., New Variational Principle for Incompressible Non-Newtonian Flow, The Physics of Fluids, **3**, 539–541 (1960).
6 Bizzell, G. P., Slattery, J. C., Non-Newtonian Boundary-Layer Flow, Chem. Eng. Sci. **17**, 777–782 (1962).
7 Богданова, В. В., Универсальные уравнения теории пространственного пограничного слоя, Мех. жидк. и газа, **6** (1968).
8 Cebeci, T., Clauss, J. M., Dyer, J., On the solution of boundary-layer equations for a non-Newtonian power-law fluid, Chem. Eng. Sci. **24**, 1031–1034 (1969).
9 Chan, Y. Y., Loitsianskii's method for boundary layers with suction and injection, AIAA J., **7**, 562–563 (1969).
10 Djordjevic, V., Generalizacija nekih poznatih metoda proračuna graničnog sloja njutnovskih tečnosti na slučaj graničnog sloja nenjutnovskih tečnosti, Matematički Vesnik, **6**, 277–281 (1969).
11 Дорфман, А. Ш., Вишневский, В. К., Приближеное решение уравнений динамического и теплового пограничных слоев для неньютоновских жидкостей при произвольных градиентах давления и температуры поверхности, Теплофизика и Теплотехника, **18**, 89—95 (1970).
12 Дорфман, А. Ш., Вишневский, В. К., Пограничный слой неньютоновских степенных жидкостей при произвольных градиентах давления, Инж-Физ. Жур. **20**, 398—404 (1971).
13 Frederickson, A. G., Bird, R. B., Non-Newtonian Flow in Annuli, Ind. Eng. Chem. **50**, 347 (1958).
14 Görtler, H., A new series for the calculation of steady laminar boundary layer flows, J. Math. Mech. **6**, 1–66 (1957).
15 Gupta, R. C., Behaviour of power-law fluids in the inlet region of a straight channel, Int. J. Non-Linear Mechanics, **5**, 325–334 (1970).
16 Gutfinger, Ch., Shinnar, R., A. I. Ch. E. Jour. **12**, 631 (1964).
17 Hartree, D. R., On an equation occuring in Falkner and Skan's approximate treatment of the equations of the boundary layer, Proc. Camb. Phil. Soc. **33**, 223–239 (1937).
18 Howarth, L., On the solution of the laminar boundary layer equations, Proc. Roy. Soc., London, A **164**, 547–579 (1938).
19 Kapur, J. N., Phys. Soc. Japan **17**, 1303 (1962).
20 Kapur, J. N., Gupta, R. C., Two-dimensional flow of power law fluids in the inlet length of a Straight channel II, ZAMM **44**, 277–284 (1963).
21 Kapur, J. N., Srivastava, R. C., Similar Solutions of the Boundary Layer Equations for Power Law Fluids, ZAMP **14**, 383–389 (1963).
22 Капустянский, С. М., Ламинарный пограничный слои в газовом потоке больших скоростей, Труды ЛПИ, **248**, 59—64 (1965).
23 Карякин, В. Е., Применение метода характеристик к решению универсальных уравнений ламинарного пограничного слоя, Аэротермодинамика, Труды ЛПИ, **313**, 20—28 (1970).
24 Lee, S. Y., Ames, W. F., Similarity solutions for Non-Newtonian Fluids, A. I. Ch. E. Jour. **12**, 700–708 (1966).
25 Leigh, D. C. F., The laminar boundary layer equation: A method of solution by means of an automatic computer, Proc. Camb. Phil. Soc. **51**, 320–332 (1955).
26 Lemieux, P. F., Dubey, R. N., Unny, T. E., Variational Method for a Pseudoplastic Fluid in a Laminar Boundary Layer Over a Flat Plate, J. App. Mech. June, 345–350 (1971).
27 Lemieux, P. F., Unny, T. E., Dubey, R. N., Variational Principle in fluid dynamics, J. Eng. Mech. Proc. Amer. Soc. Civil Eng. December, 1031–1038 (1970).
28 Лойцянский, Л. Г., Универсальные уравнения и параметрические приближения в теории ламинарного пограничного слоя, ПММ, **29**, 70—87 (1965).
29 Любенов, В. Й., Двухпараметрическое решение уравнений ламинарного пограничного слоя в газе, Аэротермодинамика, Труды ЛПИ, **313**, 28—35 (1970).
30 Metzner, A. B., Reed, J. C., Flow of Non-Newtonian Fluids-Correlation of the Laminar, Transition, and Turbulent Flow Regions, A. I. Ch., E. J. **1**, 434 (1955).
31 Reiner, M., Deformation and flow, Lewis, London (1949).
32 Rott, W., Laminare Strömung dilatanter Flüssigkeiten in der Nähe einer plötzlich in Bewegung gesetzten Platte, Chem. Eng. Sci. **26**, 739–742 (1971).
33 Сальников, В., Обобщение универсального уравнения теорий ламинарного пограничного слоя Л. Г. Лойцянского, Public. de l'Inst. Math. **13**, 105–116 (1972).
34 Schlichting, H., Boundary-Layer Theory, McGraw-Hill, Sixth. ed., New York 1968.
35 Schowalter, W. R., The Application of Boundary Layer Theory to Power-Law Pseudoplastic Fluids; Similar Solutions, A. I. Ch. E. J. **6**, 24–28 (1960).
36 Serth, R. W., Kiser, K. M., A solution of the two-dimensional boundary-layer equations for an Ostwald de Waele fluid, Chem. Eng. Sci. **22**, 945–956 (1967).
37 Slattery, J. C., Bird, R. B., Non-Newtonian flow past a sphere, Chem. Eng. Sci. **16**, 231–241 (1961).
38 Slattery, J. C., Chem. Eng. Sci. **17**, 689 (1961).
39 Shah, M. J., Petersen, E. E., Acrivos, A., Heat transfer from a cylinder to a power-law Non-Newtonian Fluid, A. I. Ch. E. J. **8**, 542–549 (1962).
40 Шульман, З. П., Берковский, Б. М., Пограничный слой неньютоновских жидкостей, Наука и Тех., Минск (1966).
41 Terril, R. M., Laminar boundary-layer flow near Separation with and without suction, Phylos. Trans. Roy. Soc. London **253**, 55–100 (1960).
42 Wilkinson, W. L., Non-Newtonian Fluids, Pergamon Press (1960).
43 Лойцянский, Л. Г., Механика жидкости и газа, Наука, Москва (1970).

Eingereicht am 15. 11. 1972, rev. Fassung: 22. 3. 1973

Anschrift: Djordje Djukić, Jovana Boskoviča 16, 21000 Novi Sad, Yugoslavia

ZAMM 54, 19 – 25 (1974)

E. MIERSEMANN

Lineare elliptische Zustände mit nichtlinearem Zielfunktional

Wir untersuchen elliptische Zustände, die durch das Dirichletproblem gegeben sind. Das Zielfunktional ist ein allgemeiner nichtlinearer Ausdruck. Wir zeigen die Existenz der optimalen Steuerung. Für eine spezielle beschränkte Steuerungsmenge beweisen wir die Regularität der optimalen Steuerung.

We study the optimal control of a system governed by the Dirichlet problem. The cost function is a general nonlinear functional. We proof the existence of the optimal control and the regularity of the optimal control for a special bounded set of controls.

Исследуются эллиптические состояния, заданные задачей Дирихле. Целевая функция является общим нелинейным выражением. Показано, что существует оптимальное управление. Для особенного, ограниченного множества управлений доказывается регулярность оптимального управления.

1. Einleitung[1])

Wir betrachten lineare elliptische Zustände mit verteiltem Parameter über einem beschränkten Grundgebiet $\mathfrak{D}$ des R^n:

$$L\,y(u) = f + u\,, \qquad u \in L_2(\mathfrak{D})\,, \qquad y(u) \in \overset{\circ}{H}_m(\mathfrak{D}) \cap H_{2m}(\mathfrak{D})\,.$$

f ist ein festes Element des $L_2(\mathfrak{D})$, und L ist ein stark elliptischer Differentialoperator $2\,m$-ter Ordnung. Der Nullraum von L kann nichttrivial sein. Das zu minimierende Funktional ist ein allgemeiner nichtlinearer Ausdruck der folgenden Art:

$$I(u) = \int_{\mathfrak{D}} F\big(x, y(u), \ldots, D^l y(u)\big)\, dx = \text{Minimum!}$$

Dabei ist $F \in C$ in den Argumenten und $l \leqq 2\,m$.

Wir betrachten auch den Fall, daß die Steuerungsvariable u explizit im Zielfunktional vorkommt.

Man hat bisher vorwiegend Steuerungsprobleme mit quadratischem Zielfunktional betrachtet [5]. Das folgende Beispiel ist dafür charakteristisch: Die Auslenkung y einer am Rand festgehaltenen Membran ist durch

$$-\,\Delta\, y(u) = f + u\,, \qquad y|_{\partial\mathfrak{D}} = 0$$

gegeben. $\mathfrak{D}$ ist ein beschränktes Gebiet des R^2. Δ ist der LAPLACE-Operator im R^2, $y(u)$ die Auslenkung der Membran, und $f + u$ ist die Kraft pro Flächeneinheit, die senkrecht auf die Membran wirkt.

Durchläuft nun u einen zulässigen Steuerungsbereich U, dann kann man i. allg. zu einer vorgegebenen Funktion $z(x_1, x_2)$ keine zulässige Steuerung $u(x_1, x_2)$ finden, so daß $y(u) = z$ gilt. Man stellt dann etwa das Minimumproblem

$$I(u) = \int_{\mathfrak{D}} |y(u) - z|^2\, dx = \text{Minimum!}$$

über der Steuerungsmenge U. Das Zielfunktional $I(u)$ ist in y quadratisch.

Wir können nun allgemeinere Zielfunktionale betrachten wie z. B.

$$I(u) = \int_{\mathfrak{D}} |y(u) - z|^p\, dx = \text{Minimum!}$$

p ist eine gerade natürliche Zahl. Für die Existenzsätze können wir für p alle positiven reellen Zahlen zulassen oder wir können auch das Minimumproblem

$$I(u) = \operatorname*{Max}_{x \in \mathfrak{D}} |y(u) - z| = \text{Minimum!}$$

über dem Steuerungsbereich U stellen.

Wir leiten im ersten Teil Existenzsätze für die optimale Steuerung her. Dabei benützen wir in Anlehnung an H. BECKERT [2] wesentlich die Coercive-Ungleichung für stark elliptische Differentialoperatoren.

Im Anschluß leiten wir das Maximumprinzip als notwendige Bedingung für die Optimalität der Steuerung her. Es ergibt sich in der üblichen Schlußweise (vgl. LIONS [5]).

Im dritten Teil betrachten wir speziell den Steuerungsbereich

$$U = \{u \mid \|u\|_{L_2} \leqq 1\}\,.$$

Die optimale Steuerung erweist sich für diese Steuerungsmenge als unendlich oft differenzierbar, wenn für die Koeffizienten von L, für F, f und für $\partial\mathfrak{D}$ dasselbe zutrifft. Für den Regularitätsbeweis benützen wir Untersuchungen in [2] über die Regularität der Lösungen von gemischtfreien Variationsproblemen.

[1]) Bezüglich der verwendeten Bezeichnungen s. Abschn. 2.

2. Bezeichnungen

$\mathcal{D}$ ist ein beschränktes und genügend reguläres Gebiet des R^n.

$C_0^\infty(\mathcal{D})$ Klasse aller C^∞-Funktionen mit in $\mathcal{D}$ kompaktem Träger.

$L_p(\mathcal{D})$ BANACHraum der zur Potenz p, $1 < p < \infty$, absolut integrablen Funktionen mit der Norm

$$||u||_{L_p} = \Big(\int_{\mathcal{D}} |u|^p \, dx\Big)^{1/p} .$$

$H_{m,p}(\mathcal{D})$ BANACHraum der Funktionen, die auf $\mathcal{D}$ bis zur Ordnung m verallgemeinerte Ableitungen besitzen, die zu L_p gehören. Die Norm ist

$$||u||_{m,p} = \Big(\int_{\mathcal{D}} \sum_{0 \leq |\mu| \leq m} |D^\mu u|^p \, dx\Big)^{1/p} .$$

Dabei haben wir die bekannte Multiindizesschreibweise benützt:

$$D^\mu = \frac{\partial^{|\mu|}}{\partial x_1^{\mu_1} \cdots \partial x_n^{\mu_n}}, \qquad |\mu| = \mu_1 + \mu_2 + \cdots + \mu_n .$$

$H_m(\mathcal{D})$ ist der Raum $H_{m,p}$ mit $p = 2$. Die Norm sei $|| \cdot ||_m$.

$\overset{0}{H}_m(\mathcal{D})$ ist der Abschluß von $C_0^\infty(\mathcal{D})$ in der H_m-Norm.

$N(L)$ Nullraum des Operators L.

$N(L)^\perp$ Orthogonalkomplement zu $N(L)$.

L^+ ist der zu L adjungierte Operator.

$C^{l,\nu}(\mathcal{D})$ Klasse der l mal stetig differenzierbaren Funktionen, deren l-te Ableitungen HÖLDERstetig sind mit dem Exponenten ν.

3. Existenzsätze

Vorgegeben sei ein stark elliptischer Differentialoperator mit genügend regulären Koeffizienten $a_j(x)$, die auf einem Gebiet $\mathcal{D}$ des R^n definiert sind:

$$L = \sum_{0 \leq |j| \leq 2m} a_j(x) \, D^j .$$

$\mathcal{D}$ ist ein beschränktes und genügend glattes Gebiet. L heißt stark elliptisch, wenn gilt

$$\sum_{|\alpha| = 2m} a_\alpha(x) \, \zeta^\alpha \geqq k \, |\zeta|^{2m} , \qquad k > 0 ,$$

mit

$$\zeta = (\zeta_1, \ldots, \zeta_n) , \qquad \zeta^\alpha = \prod_{j=1}^{n} \zeta_j^{\alpha_j} , \qquad 2m = \sum_{j=1}^{n} \alpha_j , \qquad |\zeta|^2 = \sum_{j=1}^{n} \zeta_j^2 .$$

U sei eine abgeschlossene und konvexe Teilmenge des $L_2(\mathcal{D})$.

Wir stellen das folgende Steuerungsproblem auf:

$$U_{ad} = \{u \mid u \in U \quad \text{und} \quad f + u \in N(L^+)^\perp\}:$$

$$L \, y(u) = f + u \tag{1}$$

$$u \in U_{ad} , \quad y(u) \in V \qquad \text{mit} \qquad V = \overset{0}{H}_m \cap H_{2m}$$

$$I(u) = \int_{\mathcal{D}} F\big(x, y(u), \ldots, D^l y(u)\big) \, dx = \text{Min!} , \qquad l \leqq 2m . \tag{2}$$

Aus einem bekannten SOBOLEW-Lemma folgt, daß $I(u)$ sinnvoll ist, wenn die folgenden Wachstumsbedingungen für F gelten:

$$|F| \leqq c(|y|_C) \Big(1 + \sum_{j < |k|}^{l} |D^k y|^{r_k}\Big) . \tag{3}$$

Dabei ist

$$j < 2m - \frac{n}{2} \qquad \text{und} \qquad 0 \leqq r_k \leqq \frac{2n}{n - 2(2m - k)} ,$$

$c(s)$ ist hier und im folgenden eine auf $R^+ = \{s \mid s \geqq 0\}$ stetige Funktion.

Satz 1: *Sei*

a) $l < 2m - \dfrac{n}{2}$,

b) *$F \in C$ in seinen Argumenten,*

c) *$N(L) = (0)$ oder $I(u) \to \infty$, wenn $||y||_{L_2} \to \infty$ gilt,*

d) *U beschränkt.*

Dann existiert für das Problem (1), (2) eine optimale Steuerung.

Beweis: Es gilt für alle $y \in V$ die Coercive-Ungleichung mit einer von y unabhängigen Konstanten K [1], [3]:

$$\|y\|_{2m} \leqq K(\|Ly\|_0 + \|y\|_0). \tag{4}$$

Im Falle $N(L) = (O)$ gilt [7]

$$\|y\|_{2m} \leqq K\|Ly\|_0. \tag{5}$$

Sei u_n, $u_n \in U_{ad}$, Minimalfolge für $I(u) = \text{Min!}$. Bei $N(L) = (O)$ gilt wegen (1) und (5)

$$\|y_n\|_{2m} \leqq K(\|f\|_0 + \|u_n\|_0).$$

Aus der Voraussetzung d) folgt die Beschränktheit der Folge $y_n = y(u_n)$ in der H_{2m}-Norm.

Sei $N(L) \neq (O)$, dann erhalten wir aus (1) und (4)

$$\|y_n\|_{2m} \leqq K(\|f\|_0 + \|u_n\|_0 + \|y_n\|_0).$$

Aus den Voraussetzungen c) und d) folgt dann ebenfalls die Beschränktheit der Folge y_n in der H_{2m}-Norm. Wir wählen eine schwachkonvergente Teilfolge $u_n \rightharpoonup u_0$ in L_2 und $y_n \rightharpoonup y_0$ in H_{2m} aus. Da L ein beschränkter Operator von H_{2m} in L_2 ist, gilt

$$f + u_n = Ly_n \rightharpoonup Ly_0 = f + u_0.$$

Es ist also $y_0 = y(u_0)$. Wegen des SOBOLEW-Lemmas (3)

$$|y|_{C^j} \leqq K\|y\|_{2m} \qquad \text{für} \qquad j < 2m - \frac{n}{2}$$

können wir für unsere Minimalfolge von vornherein annehmen, daß $y_n \to y_0$ in C^j gilt.

Aus den Voraussetzungen a) und b) folgt, daß der Grenzübergang unter dem Integralzeichen vollzogen werden kann:

$$\lim_{n\to\infty} \int_D F(x, y_n, \ldots, D^l y_n)\, dx = \int_D F(x, y_0, \ldots, D^l y_0)\, dx.$$

Wie Satz 1 beweist man

Satz 2: *Sei*

a) $l \leqq 2m$,
b) $\int_D F(x, \ldots, D^l y)\, dx$ *sei in* y *bezüglich der schwachen Konvergenz in* H_{2m} *stetig,*
c) $N(L) = (0)$ *oder* $I(u) \to \infty$, *wenn* $\|y\|_{L_2} \to \infty$ *gilt,*
d) U *beschränkt.*

Dann existiert eine optimale Steuerung für (1), (2).

Bemerkung: Die Stetigkeitseigenschaft b) trifft zu, wenn gilt: $l \leqq 2m - 1$ und

$$|F| \leqq c(|y|_{C^j})\left(1 + \sum_{j<k}^{l} |D^k y|^{r_k}\right)$$

mit

$$j < 2m - 1 - \frac{n}{2} \qquad \text{und} \qquad 0 \leqq r_k \leqq \frac{2n}{n - 2(2m - k - 1)}.$$

Denn es folgt aus $y_n \rightharpoonup y_0$ in H_{2m} die Konvergenz $y_n \to y_0$ in H_{2m-1}. Aus den obigen Wachstumsbedingungen folgt dann die Stetigkeit des Integrals [4].

Wir betrachten jetzt den Fall, daß die Steuerungsvariable explizit im Zielfunktional vorkommt:

$$I(u) = \int_D F(x, y(u), \ldots, D^l y(u); u)\, dx. \tag{6}$$

$I(u)$ ist sinnvoll, wenn folgende Wachstumsbedingungen gelten:

$$|F| \leqq c(|u|_{C^j})\left(1 + \sum_{j<|k|}^{l} |D^k y|^{r_k} + |u|^s\right) \tag{7}$$

mit

$$j < 2m - \frac{n}{2}, \qquad 0 \leqq r_k \leqq \frac{2n}{n - 2(2m - k)} \qquad \text{und} \qquad 0 \leqq s \leqq 2.$$

Es sei

$$F \in C^1 \quad \text{in seinen Argumenten}. \tag{8}$$

Für die partiellen Ableitungen fordern wir:

$$|F_{D^k y}| \leqq c(|y|_{C^j})\left(1 + \sum_{j<|l|} |D^l y|^{r_l} + |u|^s\right) \quad \text{für} \quad k < 2m - 1 - \frac{n}{2} \tag{9}$$

mit

$$0 \leqq r_l \leqq \frac{2n}{n - 2(2m - l - 1)} \qquad \text{und} \qquad 0 \leqq s \leqq 2.$$

Für $k \geqq 2m - 1 - \frac{n}{2}$ gelte

$$|F_{D^k y}| \leqq c(|y|_{C^j})\left(1 + \sum_{j<|l|} |D^l y|^{t_l} + |u|^s\right)$$

mit $s, t_l \geqq 0$, $\quad 0 \leqq s\, q_k \leqq 2$ und

$$0 \leqq t_l\, q_k \leqq \frac{2n}{n - 2(2m - l - 1)} .$$

Dabei ist

$$q_k = \frac{r_k}{r_k - 1} \qquad \text{und} \qquad 1 < r_k \leqq \frac{2n}{n - 2(2m - k - 1)} .$$

Lemma: *Es seien die Wachstumsbedingungen (9) erfüllt. Dann ist*

$$M(y, u) = \int_D F(x, y, \ldots, D^l y; u)\, dx$$

für beschränkte Mengen in U in der Variablen y bezüglich u gleichmäßig stetig in der starken H_{2m-1}-Konvergenz.

Beweis: Es ist

$$\int_D F(x, y_n, \ldots, D^l y_n, u)\, dx - \int_D F(x, y_0, \ldots, u)\, dx = \int_D \int_0^1 \sum_{0 \leqq |k| \leqq l} F_{D^k y}(x, y_n + t(y_0 - y_n), \ldots, u)\, D^k(y_0 - y_n)\, dt\, dx .$$

Aus der Hölderschen Ungleichung für zwei Faktoren und aus den Wachstumsbedingungen (9) folgt die Behauptung. Dabei ist noch die folgende Ungleichung [3] zu berücksichtigen:

$$||D^k y||_{L_{r_k}} \leqq K\, ||y||_{2m-1} \qquad \text{für} \qquad y \in H_{2m-1} .$$

Die Konstante K ist von y unabhängig. Es ist

$$1 < r_k \leqq \frac{2n}{n - 2(2m - k - 1)} .$$

Satz 3: *Voraussetzungen:*

a) $l \leqq 2m - 1$.
b) $F \in C^1$ *in seinen Argumenten.*
c) $N(L) = (0)$ *oder* $I(u) \to \infty$, *wenn* $||y||_{L_2} \to \infty$ *gilt.*
d) U_{ad} *ist beschränkt, oder es gilt* $I(u) \to \infty$, *wenn* $||u||_{L_2} \to \infty$ *gilt.*
e) $M(y, u) = \int_{\mathfrak{D}} F(x, y, \ldots, D^l y; u)\, dx$
ist für festes y in u unterhalbstetig bezüglich der schwachen Konvergenz in $L_2(\mathfrak{D})$.
f) *Es gelten die Wachstumsbedingungen (9).*

Dann existiert für das Probelm (1), (6) eine optimale Steuerung.

Beweis: Es sei u_n Minimalfolge für $I(u) = \text{Min!}$ Wegen der Voraussetzung d) ist $||u_n||_{L_2}$ beschränkt. Wie beim Beweis von Satz 1 schließen wir auf die Beschränktheit der Folge $||y_n||_{2m}$. Aus den Wachstumsbedingungen (7) folgt dann

$$m = \inf_{u \in U_{ad}} I(u) > -\infty .$$

Wir wählen eine schwachkonvergente Teilfolge aus, die wir wieder wie oben bezeichnen: $y_n \rightharpoonup y_0$ in H_{2m}, $u_n \rightharpoonup u_0$ in L_2. Wie beim Beweis von Satz 1 ergibt sich $y_0 = y(u_0)$. Wir zeigen

$$m = \int_{\mathfrak{D}} F(x, y_0, \ldots, D^l y_0, u_0)\, dx .$$

Es gilt $y_n \to y_0$ in H_{2m-1}.
Für beliebiges $\varepsilon > 0$ gilt, da u_n Minimalfolge ist, mit $n > n_0(\varepsilon)$

$$\int_{\mathfrak{D}} F(x, y_n, \ldots, D^l y_n, u_n)\, dx \leqq m + \frac{\varepsilon}{2} .$$

Wegen des Lemmas gilt für $n > n_1(\varepsilon)$

$$\int_{\mathfrak{D}} F(x, y_0, \ldots, D^l y_0, u_n)\, dx \leqq \int_{\mathfrak{D}} F(x, y_n, \ldots, u_n)\, dx + \frac{\varepsilon}{2} .$$

Aus beiden Ungleichungen erhalten wir für $n > \operatorname{Max}(n_0, n_1)$

$$\int_{\mathfrak{D}} F(x, y_0, \ldots, D^l y_0, u_n)\, dx \leqq m + \varepsilon .$$

Aus der Voraussetzung e) folgt die Ungleichung

$$\int_{\mathfrak{D}} F(x, y_0, \ldots, D^l y_0, u_0)\, dx \leqq \lim_{n \to \infty} \int_{\mathfrak{D}} F(x, y_0, \ldots, D^l y_0, u_n)\, dx .$$

Insgesamt erhalten wir

$$\int_{\mathfrak{D}} F(x, y_0, \ldots, D^l y_0, u_0)\, dx \leqq m + \varepsilon .$$

Da ε beliebig ist, gilt

$$m = \int_{\mathfrak{D}} F(x, y_0, \ldots, D^l y_0, u_0)\, dx .$$

Es ist also u_0 eine optimale Steuerung und $y_0 = y(u_0)$ der zugehörige Zustand.

4. Maximumprinzip

Wir leiten jetzt ganz analog zu LIONS [5] eine notwendige Bedingung für die Optimalität der Steuerung her. Wir verlangen

$$F \in C^2 \quad \text{in den Argumenten} . \tag{10}$$

Es sei u die optimale Steuerung und $v \in U_{ad}$. Für $0 \leqq \varepsilon \leqq 1$ ist $u + \varepsilon(v - u) \in U_{ad}$, da U_{ad} konvex ist. Wir entwickeln $I(u + \varepsilon(v - u))$ nach ε und erhalten

$$I(u + \varepsilon(v - u)) = I(u) + \varepsilon\, V(u, v - u) + \frac{\varepsilon^2}{2} \tilde{Q}(v - u, v - u). \tag{11}$$

Dabei ist

$$V(u, v - u) = \int_{\mathfrak{D}} \sum_{0 \leqq |j| \leqq l} F_{D^j y}(x, \ldots)\, D^j(y(v) - y(u))\, dx + \int_{\mathfrak{D}} F_u \cdot (v - u)\, dx \tag{12}$$

und

$$\frac{\varepsilon^2}{2} \tilde{Q} = \int_{\mathfrak{D}} \int_0^1 \int_0^t \Phi''(t')\, dt'\, dt\, dx \tag{13}$$

mit

$$\Phi(t) = F(x, y(u) + t\,\varepsilon(y(v) - y(u)), \ldots, u + t\,\varepsilon(v - u)) .$$

Es sei $l < 2m - \frac{n}{2}$ erfüllt. Für das Integral (1) sind V und $\tilde{Q}$ beschränkt. Für das Integral (6) ist V beschränkt, wenn gilt

$$\left.\begin{aligned} |F_{D^j y}| &\leqq c(|y|_{C^l})\,(1 + |u|^{k_j}) , & 0 &\leqq k_j \leqq 2 , \\ |F_u| &\leqq c(|y|_{C^l})\,(1 + |u|^{k}) , & 0 &\leqq k \leqq 1 . \end{aligned}\right\} \tag{14}$$

Das Integral $\tilde{Q}$ ist beschränkt, wenn gilt

$$\begin{aligned} |F_{D^j y\, D^j y}| &\leqq c(|y|_{C^l})\,(1 + |u|^{k_{jj}}) , & 0 &\leqq k_{ij} \leqq 2 , \\ |F_{D^j y, u}| &\leqq c(|y|_{C^l})\,(1 + |u|^{k_j}) , & 0 &\leqq k_i \leqq 1 , \\ |F_{uu}| &\leqq c(|y|_{C^l}) . \end{aligned} \tag{15}$$

Bemerkung: Es lassen sich auch im Falle $l \geqq 2m - \frac{n}{2}$ hinreichende Bedingungen für die Beschränktheit von I, V und $\tilde{Q}$ angeben, [2], [4].

Es gilt, da u optimal ist, für alle $v \in U_{ad}$

$$I(u + \varepsilon(v - u)) \geqq I(u) , \qquad 0 < \varepsilon < 1 .$$

Daraus folgt wegen der Entwicklung (11) für alle $v \in U_{ad}$

$$V(u, v - u) = \int_{\mathfrak{D}} \sum_{0 \leqq |j| \leqq l} F_{D^j y}(x, \ldots)\, D^j(y(v) - y(u))\, dx + \int_{\mathfrak{D}} F_u \cdot (v - u)\, dx \geqq 0 .$$

$w = y(v) - y(u)$ ist Lösung von $Lw = v - u$. Wenn $N(L) \neq (0)$ gilt, dann ist w nicht eindeutig bestimmt.

Folgerung: Ist die Steuerung u, der Zustand $y(u)$ in C^l und $F \in C^{l+1}$ und gilt $N(L^+) = (0)$, dann kann man die adjungierte Steuerung $p(u)$ definieren:

$$L^+ p(u) = \sum_{0 \leqq |i| \leqq l} (-1)^{|j|} D^j F_{D^j y}(x, y(u), \ldots, u) , \qquad p(u) \in \overset{0}{H}_m .$$

Damit gilt dann für alle $v \in U_{ad}$

$$\int_{\mathfrak{D}} (p(u) + F_u)\,(v - u)\, dx \geqq 0 .$$

Bemerkung: Bei $N(L^+) \neq (0)$ ist die Existenz des adjungierten Zustandes $p(u)$ nicht gesichert.

5. Regularität

Wir betrachten den folgenden Steuerungsbereich:

$$U = \{u \mid ||u||_{L_2} \leqq 1\} .$$

Satz 4: *Voraussetzungen:*

a) $l < 2m - \frac{n}{2}$.

b) $N(L) = (0)$.

c) *Die optimale Steuerung sei keine Extremale, d. h. für ein* $v \in L_2(\mathfrak{D})$ *gilt* $V(u, v) \neq 0$.

d) $F \in C^\infty$ *in den Argumenten,* $f \in C^\infty(\overline{\mathfrak{D}})$, *und es sei der Rand von* $\mathfrak{D}$ *unendlich oft differenzierbar.*

Dann gilt $u \in C^\infty(\overline{\mathfrak{D}})$ *für die optimale Steuerung* u *von* (1), (2).

Beweis: Für die optimale Steuerung u gilt $||u||_{L_2} = 1$. Aus $||u||_{L_2} < 1$ folgt nämlich wegen der Entwicklung (11) $V(u, v) = 0$ für alle $v \in L_2(\mathcal{D})$. Das ist aber ein Widerspruch zur Voraussetzung c). Es sei $\varepsilon > 0$. Wegen der Entwicklung (11) folgt aus $||u + \varepsilon v||_{L_2} < 1$ die Ungleichung $V(u, v) \geqq 0$. Es ist

$$||u + \varepsilon v||_{L_2} < 1 \leftrightarrow ||u + \varepsilon v||^2_{L_2} - ||u||^2_{L_2} < 0 ,$$

d. h. aus $2(u, v) < 0$ folgt $V(u, v) > 0$. Nach einem Lemma in [2] existiert eine negative Konstante λ mit

$$2\,\lambda(u, v) = V(u, v) \qquad \text{für alle} \quad v \in L_2(\mathcal{D}) .$$

Es ist also

$$2\,\lambda(u, v) = \int_{\mathcal{D}} \sum_{0 \leqq |j| \leqq l} F_{D^j y}\, D^j z(x)\, dx .$$

Wir führen die GREENsche Funktion zu L ein, die wegen der Voraussetzung b) existiert,

$$z(x) = \int_{\mathcal{D}} G(x, x')\, v(x')\, dx' .$$

Wir erhalten für alle $v \in L_2(\mathcal{D})$:

$$2\,\lambda(u, v) = \int_{\mathcal{D}} \int_{\mathcal{D}} \sum_{0 \leqq |j| \leqq l} F_{D^j y}(x, \ldots)\, D_x^j\, G(x, x')\, v(x')\, dx'\, dx .$$

Daraus folgt

$$2\,\lambda\, u(x') = \int_{\mathcal{D}} \sum_{0 \leqq |j| \leqq l} F_{D^j y}(x, \ldots)\, D_x^j\, G(x, x')\, dx . \tag{16}$$

Wie in [2] folgt wegen der Voraussetzung a)

$$u \in C^{m, \mu}(\overline{\mathcal{D}}) .$$

Aus (1) folgt dann [1]

$$y(u) \in C^{3m, \mu}(\overline{\mathcal{D}}) .$$

Da nach Voraussetzung a) $2\,m > l + 1$ gilt, können wir die rechte Seite von (16) mindestens einmal partiell integrieren. Dadurch verringert sich die Ableitungsordnung bei der GREENschen Funktion um eins. Die Regularität von u erhöht sich iterativ [2]. Wir erhalten schließlich $u \in C^\infty(\overline{\mathcal{D}})$.

Bemerkungen:

1. Die Voraussetzung c) ist in folgendem Beispiel erfüllt. Es sei

$$L\, y(u) = f + u \qquad \text{und} \qquad I(u) = \int_{\mathcal{D}} |y(u) - h|^2\, dx .$$

h ist ein festes Element aus $L_2(\mathcal{D})$ und $h \notin \overset{0}{H}_m$. Wäre u Extremale, dann würde $y(u) = h$ gelten. Das kann aber nicht sein, da $y(u) \in \overset{0}{H}_m$ gilt.

2. Werden Wachstumsbedingungen an die $F_{D^j y}$ gestellt, dann kann auch für l: $2\,m - \frac{n}{2} \leqq l \leqq 2\,m - 1$, $u \in C^\infty$ nachgewiesen werden, indem man die Ergebnisse von [2] auf das obige Steuerungsproblem anwendet.

Wir leiten jetzt einen Regularitätssatz für das Problem (1), (6) her.

Satz 5: *Voraussetzungen:*

a) *Das Problem* (1), (6) *sei über U lösbar.*
b) *Für die optimale Steuerung u und den zugehörigen Zustand $y(u)$ soll gelten*

$$F_{uu}(x, y(u), \ldots, D^l\, y(u)) \geqq 0 .$$

c) *Für* $0 \leqq r \leqq 2$*:*

$$|F| \leqq c(|y|_{C^l})\, (1 + |u|^r) .$$

d) *Mit* $0 \leqq r_k < 2$ *verlangen wir*

$$|F_{D^k y}| \leqq c(|y|_{C^l})\, (1 + |u|^{r_k}) .$$

e) $$l < 2\,m - \frac{n}{2} \operatorname*{Max}_k r_k \qquad \text{und} \qquad l < 2\,m - \frac{n}{2} .$$

Weiterhin sollen die Voraussetzungen a)—d) von Satz 4 gelten.
Dann ist $u \in C^\infty(\mathcal{S})$ für jedes in $\mathcal{D}$ kompakte Gebiet $\mathcal{S}$.

Beweis: Die Darstellungsformel (16) hat jetzt die Form

$$2\,\lambda\, u(x') = \int_{\mathcal{D}} \sum_{0 \leqq |j| \leqq l} F_{D^j y}(x, \ldots)\, D_x^j\, G(x, x')\, dx + F_u . \tag{17}$$

Wegen d) gilt $F_{D^j y} \in L_{p_j}$ mit $p_j = 2/r_j$. Auf Grund der Singularitätsordnung von $D_x^j\, G(x, x')$ gilt für jedes in $\mathcal{D}$ kompakte Gebiet $\mathcal{S}$ ([6], S. 73, Theorem 3.4.1)

$$\int_{\mathcal{D}} F_{D^j y}(x, \ldots)\, D_x^j\, G(x, x')\, dx \in H_{2m - |j|, p_j}(\mathcal{S}) .$$

Folglich gilt mit $p = \operatorname*{Min}_j p_j$

$$2\,\lambda\, u - F_u \in H_{2m - l, p} .$$

Aus (17) erkennen wir wegen der Voraussetzung e), daß $2\lambda u - F_u \in C^\mu(\mathcal{S})$ gilt. Aus der zweiten Bedingung von e) folgt $y(u) \in C^{1,\mu}(\mathcal{S})$.

Wegen b) ist

$$(2\lambda u - F_u)_u \leqq 2\lambda .$$

Wir können also $2\lambda u - F_u$ nach u auflösen, und wir erhalten $u \in C^\mu(\mathcal{S})$. Daraus folgt [1] $y(u) \in C^{2m,\mu}(\mathcal{S})$. Wie bei Satz 4 erhalten wir dann aus der Darstellungsformel (17) $u \in C^\infty(\mathcal{S})$.

Literatur

1 AGMON, S., DOUGLIS, A. and NIRENBERG, L., Estimates near the boundary for solutions of elliptic partial differential equations satisfying general boundary conditions, I. Comm. on Pure and Appl. Math. **12**, 623–727 (1959).
2 BECKERT, H., Über eine bemerkenswerte Klasse gemischtfreier Variationsprobleme höherer Ordnung, Math. Zeitschr. **103**, 18–29 (1968).
3 BROWDER, F. E., On the spectral theory of elliptic differential operators, I. Math. Ann. **142**, 22–130 (1961).
4 BROWDER, F. E., Nonlinear elliptic boundary value problems II, Trans. Amer. Math. Soc. **117**, 530–550 (1961).
5 LIONS, J. L., Optimal control of systems governed by partial differential equations, Springer 1971.
6 MORREY, C. B., Multiple integrals in the calculus of variations, Springer 1966.
7 SCHECHTER, M., Solutions of the Dirichlet problem for systems not necessarily strongly elliptic equations, Comm. on Pure and Appl. Math. **12**, 241–247 (1959).

Eingereicht am 10. 1. 1973.

Anschrift: Dr. ERICH MIERSEMANN, 701 Leipzig, Karl-Marx-Universität, Sektion Mathematik, Karl-Marx-Platz, DDR

ZAMM 54, 27 – 37 (1974)

N. LIRON / H. E. WILHELM

Integration of the Magnetohydrodynamic Boundary-Layer Equations by Meksyn's Method*)

Es wird das nichtlineare Randwertproblem formuliert, das die räumliche Entwicklung der Grenzschicht in einer leitenden Flüssigkeit beschreibt, die bei Vorhandensein eines äußeren Druckgradienten quer zu einem inhomogenen Magnetfeld entlang einer ebenen Platte strömt. Mit Hilfe von Transformationen, Reihenentwicklungen und Integrationsverfahren (Sattelpunkt-Methode), wie sie von Meksyn in die Analysis hydrodynamischer Grenzschichten eingeführt worden sind, werden analytische Lösungen für das Geschwindigkeitsfeld sowohl in Punkten der anliegenden als auch der abgelösten magnetohydrodynamischen Strömung gewonnen. Der Anwendung auf magnetohydrodynamische Diffusor-Strömungen wird besondere Aufmerksamkeit geschenkt. Es wird gezeigt, daß ein inhomogenes Magnetfeld die Grenzschichtablösung unterdrücken kann.

The nonlinear boundary-value problem describing the spatial development of the boundary-layer in a conducting fluid flowing transverse to an inhomogeneous magnetic field along a flat plate with external pressure gradient is formulated. Analytical solutions of the velocity field are obtained for unseparated and separated points in the magnetohydrodynamic flow by means of transformations, series expansions, and integration procedures (method of steepest descent) introduced by Meksyn in the analysis of ordinary boundary layers. In an application, attention is given to magnetohydrodynamic diffuser flows. It is shown that an inhomogeneous magnetic field can suppress boundary-layer separation.

Формулируется нелинейная задача на краевые значения, описывающая пространственное развитие пограничного слоя в проводящей жидкости, текущей вдоль плоской пластины под влиянием внешнего градиента давления, направленного поперёк к неоднородному магнитному полю. С помощью трансформаций, развития в ряды и специальных приёмов интегрирования (напр.: метода скорейшего спуска), которые были введены в анализ пограничных слоев Мэксыном, определяются аналитические решения для поля скоростей в точках прилегания и отслаивания магнито-гидродинамического потока. Особое внимание уделяется применению к магнито-гидродинамическому диффузионному течению. Показано, что неоднородное магнитное поле способно подавить отслаивание пограничного слоя.

Introduction

Magnetohydrodynamic boundary-layers are observed in various technical systems employing liquid metal and plasma flows transvers to magnetic fields [1], [2]. The theoretical work on magnetohydrodynamic boundary layers has been concerned mainly with extensions of the standard methods of boundary-layer theory for non-conducting fluids [3]. One of the more elegant mathematical approaches of classical boundary-layer theory is due to MEKSYN [4], and seems to have been applied to magnetohydrodynamic boundary-layers only in a few special cases and in relatively low analytical approximations, respectively [5], [6].

In the following, MEKSYN's method is extended to the general magnetohydrodynamic boundary layer equations for a conducting flow along a plate and transverse to an inhomogeneous magnetic field, in presence of an external pressure gradient. Series solutions are obtained for i) unseparated and ii) separated points in the magnetohydrodynamic flow, which reduce to those of MEKSYN for *the general equation* [7] if the magnetic field is set to zero. The HALL-effect is disregarded, i. e., it is assumed that the gyration frequency, $\omega = |e|\, B/m$, and the collision time, τ, of the electrons satisfy the inequality,

$$\omega \tau \ll 1 .$$

MHD Boundary-Layer Problem

Let cartesian coordinates (x, y, z) be introduced for the description of the magnetohydrodynamic flow (Fig. 1). The semi-infinite plate $(0 \leqq x \leqq \infty)$ is located in the plane $y = 0$, and the oncoming flow is inhomogeneous (pressure gradient) and parallel to the x-axis

$$\boldsymbol{u}_\infty = \big(u_\infty(x), 0, 0\big) . \tag{1}$$

The external magnetic field is parallel to the x-y plane, and is assumed to be given in the general form

$$\boldsymbol{B} = \big(B_x(x, y)\, ,\ B_y(x, y), 0\big) . \tag{2}$$

The components $B_x(x, y)$ and $B_y(x, y)$ are determined by MAXWELL's equations of the vacuum

$$\frac{\partial B_x}{\partial x} + \frac{\partial B_y}{\partial y} = 0\, , \qquad \frac{\partial B_x}{\partial y} - \frac{\partial B_y}{\partial x} = 0 \tag{3}$$

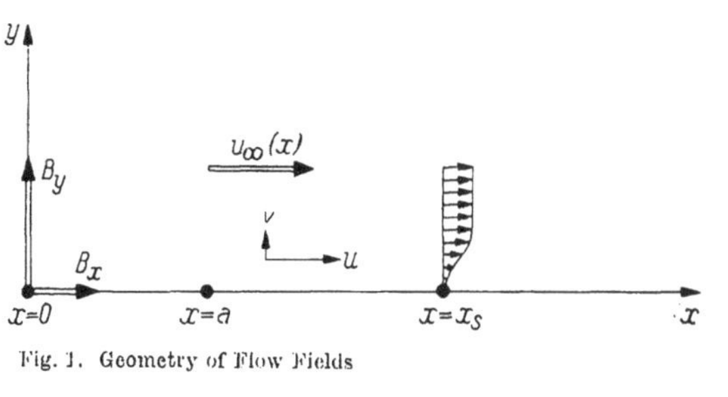

Fig. 1. Geometry of Flow Fields

*) Supported in part by the U. S. Office of Naval Research.

and the respective boundary conditions imposed by the system generating the magnetic field. The induced magnetic field is assumed to be small compared to the external magnetic field. This implies a small magnetic REYNOLDS number for the semi-infinite plate ($0 \leqq x \leqq \infty$) and the free stream velocity $u_\infty(x)$ [μ_0 = permeability of vacuum, σ = electrical conductivity],

$$R_B = \mu_0 \, \sigma \, u_\infty(x) \, x \ll 1 \,, \quad u(x)_\infty \, x \to C \,, \quad 0 \leqq C < \infty \,, \qquad \text{for} \quad x \to \infty \,. \tag{4}$$

Because of the small width δ of the boundary-layer it is for any scalar flow field $g(x, y)$,

$$|\partial g/\partial x| \ll |\partial g/\partial y| \,,$$

except for the pressure field $p(x, y) = p(x)$ [Eq. (11)], and the magnetic field $\boldsymbol{B}(x, y)$, which is independent of the boundary layer. The velocity components of the flow in the x and y directions are of the order $u \sim u_\infty$ and $v \sim u_\infty \, \delta/L$. It follows, that the boundary-layer over the flat plate with pressure gradient is described by the nonlinear boundary-value problem:

$$u \frac{\partial u}{\partial x} + v \frac{\partial u}{\partial y} = u_\infty \frac{du_\infty}{dx} + \frac{\sigma}{\varrho} (u_\infty - u) \, B_y^2 + \nu \frac{\partial^2 u}{\partial y^2} \,, \tag{5}$$

$$\frac{\partial u}{\partial x} + \frac{\partial v}{\partial y} = 0 \,, \tag{6}$$

where

$$u(x, y)_{y=0} = 0 \,, \tag{7}$$

$$v(x, y)_{y=0} = 0 \,, \tag{8}$$

$$u(x, y)_{y=\infty} = u_\infty(x) \,, \tag{9}$$

$$v(x, y)_{y=\infty} = 0 \,, \tag{10}$$

specify that the velocity components vanish at the plate [Eqs. (7)—(8)] and the longitudinal [Eq. (9)] and transverse [Eq. (10)] velocity components approach $u_\infty(x) \neq 0$ and $v_\infty(x) = 0$, respectively, at $y = \infty$. The equation of motion in the y-direction indicates that

$$\partial p/\partial y = 0 \,, \quad \text{i. e.: } p(x, y) = p_\infty(x) \,. \tag{11}$$

In Eq. (5), the pressure gradient has been eliminated by means of the momentum equation of the oncoming flow,

$$-\frac{1}{\varrho} \frac{dp}{dx} = -\frac{1}{\varrho} \frac{dp_\infty}{dx} = u_\infty \frac{du_\infty}{dx} + \frac{\sigma}{\varrho} (E_z + u_\infty \, B_{y\infty}) \, B_{y\infty} \,, \tag{12}$$

$$E_z = E_0 \,.$$

The y-dependence of the magnetic flied in Eq. (5) can be disregarded within the boundary layer approximation, since $B_y(x, y) \approx B_y(x, \delta) \approx B_{y\infty}$ for $0 \leqq y \leqq \delta$. Because of $\nabla \times \boldsymbol{E} = 0$ and $\nabla \cdot \boldsymbol{j} = \sigma \, \partial \, (E_z + u \, B_y)/\partial z = 0$, the electric field is homogeneous, E_0 = constant. It is determined by the respective boundary conditions of the flow in the z-direction [8].

Upon introduction of a stream function $\psi(x, y)$ consistent with the continuity Eq. (6), and the substitutions [7]

$$\psi = (2 \, \nu \, u_0 \, \xi)^{1/2} \, f(\xi, \eta) \,, \tag{13}$$

$$\xi = \int_{x_0}^{x} [u_\infty(x')/u_0] \, dx' \,, \quad \eta = (2 \, \nu \, u_0 \, \xi)^{-1/2} \, u_\infty(x) \, y \,, \tag{14}$$

which give

$$u = \frac{\partial \psi}{\partial y} = u_\infty(\xi) \frac{\partial f}{\partial \eta} \,, \tag{15}$$

$$v = -\frac{\partial \psi}{\partial x} = -u_\infty(\xi) \sqrt{\nu} \, (2 \, u_0 \, \xi)^{-1/2} \left\{ f + 2 \, \xi \frac{\partial f}{\partial \xi} - [1 + \lambda(\xi)] \, \eta \frac{\partial f}{\partial \eta} \right\} , \tag{16}$$

the boundary-value problem defined in Eqs. (5)—(10) is transformed to [$f' \equiv \partial f/\partial \eta$]:

$$f''' + f f'' = \lambda(\xi) \, (1 - f'^2) + \beta(\xi) \, (1 - f') + 2 \, \xi \left(f' \frac{\partial f'}{\partial \xi} - f'' \frac{\partial f}{\partial \xi} \right) \tag{17}$$

where

$$f(\xi, \eta)_{\eta=0} = 0 \tag{18}$$

$$f'(\xi, \eta)_{\eta=0} = 0 \tag{19}$$

$$f'(\xi, \eta)_{\eta=\infty} = 1 \tag{20}$$

and

$$\lambda(\xi) \equiv -2\,\xi\, d \ln u_\infty(\xi)/d\xi\,, \tag{21}$$

$$\beta(\xi) \equiv -\sigma\, u_0\, \xi\, B_y^2(\xi, \eta_\delta) \Big/ \frac{1}{2} \varrho\, u_\infty^2(\xi)\,. \tag{22}$$

The variation of $B_y(\xi, \eta)$ with η is negligible for $0 \leqq \eta \leqq \eta(\delta)$ i. e., $B_y(\xi, \eta) = B_y(\xi, \eta_\delta)$ in Eq. (22). u_0 is a constant of the dimension of a velocity of the order of $u_\infty(\xi)$

Following MEKSYN [7], Eq. (17) is formally written as an inhomogeneous equation for $f'' = \partial^2 f/\partial\eta^2$:

$$\frac{\partial}{\partial\eta} f'' + f f'' = P(\xi, \eta) \tag{23}$$

where

$$P(\xi, \eta) = \lambda[1 - f'^2] + \beta(1 - f') + 2\,\xi \left(f' \frac{\partial f'}{\partial \xi} - f'' \frac{\partial f}{\partial \xi} \right); \tag{24}$$

hence

$$\frac{\partial^2 f}{\partial\eta^2} = \varphi(\xi, \eta)\, e^{-F(\xi, \eta)} \tag{25}$$

where

$$F(\xi, \eta) \equiv \int_0^\eta f(\xi, \eta)\, d\eta \tag{26}$$

and

$$\varphi(\xi, \eta) \equiv c(\xi) + \int_0^\eta P(\xi, \eta)\, e^{F(\xi, \eta)}\, d\eta \tag{27}$$

is a slowly varying function of η for any ξ as can be shown from the asymptotic solution for $f(\xi, \eta)$.

The boundary-value problem in Eqs. (17)—(20) is solved by the following method [7]:

i) $f(\xi, \eta)$ is expanded in powers of η, the coefficients $a_n(\xi)$ being functions of ξ,

$$f(\xi, \eta) = \sum_{n=2}^{\infty} a_n(\xi) \frac{\eta^n}{n!}\,, \tag{28}$$

where $a_0 = 0$ and $a_1 = 0$ in view of the boundary conditions in Eqs. (18) and (19). Further, by Eq. (26)

$$F(\xi, \eta) = \sum_{n=2}^{\infty} a_n(\xi) \frac{\eta^{n+1}}{(n+1)!}\,. \tag{29}$$

j) The expansion of Eq. (28) is inserted in Eq. (17), and the coefficients $a_{n \geqq 3}(\xi)$ are found in terms of $a_2(\xi)$ and $a_2'(\xi) = da_2(\xi)/d\xi$.

k) The function $\varphi(\xi, \eta)$ defined in Eq. (27) is expanded in powers of η, the coefficients $b_n(\xi)$ being functions of ξ,

$$\varphi(\xi, \eta) = \sum_{n=0}^{\infty} b_n(\xi) \frac{\eta^n}{n!}\,. \tag{30}$$

l) Inserting the expansion for $\varphi(\xi, \eta)$ [Eq. (30)] and the expansions for $f(\xi, \eta)$ [Eq. (28)] and $F(\xi, \eta)$ [Eq. (29)] in Eq. (25), and comparing the terms on both sides, one obtains the coefficients b_n in terms of the coefficients $a_n = a_n(a_2)$.

m) Finally, Eq. (25) is integrated once to obtain $f' = \partial f/\partial\eta$, which is evaluated for $\eta = \infty$. The boundary condition in Eq. (20) gives then the equation $f'(\xi, \infty) = 1$ determining the still undetermined parameter $a_2(\xi)$.

Unseparated Flow

For an unseparated boundary-layer it is $a_2(\xi) = f''(\xi, \eta)_{\eta=0} \neq 0$. In this case, the operations i) — j) yield

$$\sum_{m=2}^{\infty} a_{m+1} \frac{\eta^{m-2}}{(m-2)!} = \gamma - \sum_{m,n=2}^{\infty} \frac{a_m}{m!} \frac{a_n}{(n-2)!} \eta^{m+n-2} - \lambda \sum_{m,n=2}^{\infty} \frac{a_m}{(m-1)!} \frac{a_n}{(n-1)!} \eta^{m+n-2} -$$
$$- \beta \sum_{m=2}^{\infty} \frac{a_m}{(m-1)!} \eta^{m-1} + 2\,\xi \sum_{m,n=2}^{\infty} \left[\frac{a_m}{(m-1)!} \frac{a_n'}{(n-1)!} - \frac{a_m}{(m-2)!} \frac{a_n'}{n!} \right] \eta^{m+n-2}\,, \tag{31}$$

where

$$\gamma \equiv \lambda + \beta\,, \qquad a_n' \equiv da_n/d\xi\,. \tag{32}$$

Hence:

$$\begin{aligned}
a_2 &= a\,, \quad a_3 = \gamma\,, \quad a_4 = -\beta\, a \\
a_5 &= -\beta\,\gamma + 2\,\xi\, a'\, a - (1 + 2\,\lambda)\, a^2 \\
a_6 &= \beta^2\, a - (4 + 6\,\lambda)\,\gamma\, a + 4\,\xi\,\gamma'\, a \\
a_7 &= -(4 + 6\,\lambda)\,\gamma^2 + 2\,\xi\,(2\,\gamma\,\gamma' - 2\,\beta\, a'\, a - 3\,\beta'\, a^2) + \beta[\beta\,\gamma + (8 + 10\,\lambda)\, a^2] \\
a_8 &= \beta[6(5 + 6\,\lambda)\,\gamma - \beta^2]\, a - 6\,\xi(3\,\gamma\,\beta' + 2\,\beta\,\gamma')\, a - 4\,\xi(3 + 8\,\lambda)\, a'\, a^2 + 4\,\xi^2\,(4\, a\, a'' - a'^2)\, a + \\
&\quad + [(11 + 10\,\lambda)\,(1 + 2\,\lambda) - 16\,\xi\,\lambda']\, a^3 \\
a_9 &= \beta[6(5 + 6\,\lambda)\,\gamma^2 - \gamma\,\beta^2 - 3(13 + 14\,\lambda)\,\beta\, a^2] + 2\,\xi\{-3(2\,\beta\,\gamma' + 3\,\beta'\,\gamma)\,\gamma + \\
&\quad + [3\,\beta^2 - \gamma\,(10 + 42\,\lambda)]\, a'\, a + [18\,\beta\,\beta' + 48\,(\beta'\,\gamma - \beta\,\gamma') - (27 + 92\,\lambda)\,\gamma']\, a^2\} + \\
&\quad + 4\,\xi^2\,[9\,\gamma\, a'^2 + 9\,\gamma\, a''\, a - 13\,\gamma'\, a'\, a + 10\,\gamma''\, a^2] + 2\,(45 + 113\,\lambda + 66\,\lambda^2)\,\gamma\, a^2\,.
\end{aligned} \tag{33}$$

Or, in general

$$a_{n+3} = \sum_{m=2}^{n} \binom{n}{m-1} a_m \left[2\,\frac{n+3-2\,m}{n+2-m}\,\xi\, a'_{n+2-m} - \left(\frac{n+1}{m} + \lambda - 1\right) a_{n+2-m} \right] - \beta\, a_{n+1}\,, \qquad n = 2, 3, 4, \ldots \tag{33'}$$

The operations k) — l), in which the functions e^F is expanded in a power series of $F = \int f\, d\eta$, yield

$$\sum_{m=0}^{\infty} b_m \frac{\eta^m}{m!} = \sum_{p=2}^{\infty} a_p \frac{\eta^{p-2}}{(p-2)!} \sum_{q=0}^{\infty} \frac{1}{q!} \left(\sum_{m=2}^{\infty} a_m \frac{\eta^{m+1}}{(m+1)!} \right)^q . \tag{34}$$

Hence:

$$\begin{aligned}
b_0 &= a\,, \quad b_1 = \gamma\,, \quad b_2 = -\beta\, a \\
b_3 &= -\beta\,\gamma + 2\,\xi\, a'\, a - 2\,\lambda\, a^2 \\
b_4 &= [(1 - 6\,\lambda)\,\gamma + \beta^2]\, a + 4\,\xi\,\gamma'\, a \\
b_5 &= (1 - 6\,\lambda)\,\gamma^2 + \beta^2\,\gamma - 2\,\xi\,(-2\,\gamma\,\gamma' + 2\,\beta\, a'\, a + 3\,\beta'\, a^2) + (-3 + 10\,\lambda)\,\beta\, a^2 \\
b_6 &= [6\,\gamma\,(-2 + 6\,\lambda) - \beta^2]\,\beta\, a - 2\,\xi\,[3(3\,\beta'\,\gamma + 2\,\beta\,\gamma')\, a - (15 - 16\,\lambda)\, a'\, a^2 + 8\,\lambda'\, a^3] + \\
&\quad + 4\,\xi^2\,(4\, a''\, a - a'^2)\, a - 10\,(1 - 2\,\lambda)\,\lambda\, a^3
\end{aligned} \tag{35}$$

where the a_n ($n \geqq 2$) have been expressed in terms of $a_2 = a$ according to Eq. (33).

In accordance with the operation m), Eq. (25) is integrated to

$$f'(\xi, \eta) = \int_0^{\eta} e^{-F(\xi,\eta)}\, \varphi(\xi, \eta)\, d\eta \tag{36}$$

where $\varphi(\xi, \eta)$ is slowly varying, and $F(\xi, \eta)$ is positive with a stationary point at $\eta = 0$. Let

$$F(\xi, \eta) = \eta^3 \sum_{m=0}^{\infty} c_m\, \eta^m = \tau\,, \qquad c_m = \frac{a_{m+2}}{(m+3)!} \tag{37}$$

and

$$\varphi(\xi, \eta) \frac{d\eta}{d\tau} = \sum_{m=0}^{\infty} d_m\, \tau^{(m-2)/3} \tag{38}$$

where the expansion starts with $\tau^{-2/3}$, since η starts with $\tau^{1/3}$ [Eq. (37)]. Accordingly [7],

$$\int^{(+0)} \frac{\varphi(\xi, \eta)\, d\eta}{\tau^{(m+1)/3}} = d_m \int^{(+0)(+0)(+0)} \frac{d\tau}{\tau} = 6\,\pi\, i\, d_m$$

i. e.

$$d_m = \frac{1}{6\,\pi\, i} \int^{(+0)} \frac{\varphi(\xi, \eta)\, d\eta}{\tau^{(m+1)/3}}\,. \tag{39}$$

From Eq. (39) it follows that d_m is the coefficient of η^{-1} in the expansion of $(1/3)\,\varphi(\eta)\,\tau^{-(m+1)/3}$ in ascending powers of η, where

$$\varphi(\xi, \eta)\,\tau^{-(m+1)/3} = \left(b_0 + \frac{b_1}{1!}\eta^1 + \frac{b_2}{2!}\eta^2 + \ldots\right)(c_0 + c_1\,\eta^1 + c_2\,\eta^2 + \ldots)^{-(m+1)/3} \tag{40}$$

hence d_m ($m = 0, 1, 2, \ldots$) is given as one-third of the coefficient of η^m of the right side of Eq. (40).

Upon substitution of Eqs. (37) and (38), the integral in Eq. (36) is evaluated in terms of the incomplete gamma functions $\gamma\left(\frac{m+1}{3}, F\right)$ [9],

$$f'(\xi,\eta) = \sum_{m=0}^{\infty} d_m \int_0^{F(\xi,\eta)} e^{-\tau}\,\tau^{(m-2)/3}\,d\tau = \sum_{m=0}^{\infty} d_m\,\gamma\left[\frac{m+1}{3}, F(\xi,\eta)\right]. \tag{41}$$

Finally, the still undetermined expansion parameter $a(\xi)$ is implicitly given by the boundary condition in Eq. (20),

$$f'(\xi,\eta)_{\eta=\infty} = \sum_{m=0}^{\infty} d_m \Gamma\left(\frac{m+1}{3}\right) = 1,\quad d_m = d_m(a, a', a'' \dots) \tag{42}$$

where the complete gamma function is defined by $\Gamma\left(\frac{m+1}{3}\right) = \gamma\left(\frac{m+1}{3}, \infty\right)$.

The procedure indicated [see Eq. (40)] gives the coefficients d_m first in terms of the b_m and c_m:

$$\begin{aligned}
d_0 &= \frac{1}{3} c_0^{-1/3}\, b_0\\
d_1 &= \frac{1}{3} c_0^{-2/3}\left(b_1 - \frac{2}{3}\frac{c_1}{c_0} b_0\right)\\
d_2 &= \frac{1}{3} c_0^{-3/3}\left[\frac{b_2}{2!} - \frac{c_1}{c_0} b_1 + b_0\left(\frac{c_1^2}{c_0^2} - \frac{c_2}{c_0}\right)\right]\\
d_3 &= \frac{1}{3} c_0^{-4/3}\left[\frac{b_3}{3!} - \frac{4}{3}\frac{c_1}{c_0}\frac{b_2}{2!} + b_1\left(\frac{14}{9}\frac{c_1^2}{c_0^2} - \frac{4}{3}\frac{c_2}{c_0}\right) + b_0\left(-\frac{140}{81}\frac{c_1^3}{c_0^3} + \frac{28}{9}\frac{c_1 c_2}{c_0^2} - \frac{4}{3}\frac{c_3}{c_0}\right)\right]\\
d_4 &= \frac{1}{3} c_0^{-5/3}\left[\frac{b_4}{4!} - \frac{5}{3}\frac{c_1}{c_0}\frac{b_3}{3!} + \frac{b_2}{2!}\left(\frac{20}{9}\frac{c_1^2}{c_0^2} - \frac{5}{3}\frac{c_2}{c_0}\right) + b_1\left(-\frac{220}{81}\frac{c_1^3}{c_0^3} + \frac{40}{9}\frac{c_1 c_2}{c_0^2} - \frac{5}{3}\frac{c_3}{c_0}\right) + \right.\\
&\quad \left. + b_0\left(\frac{770}{243}\frac{c_1^4}{c_0^4} - \frac{220}{27}\frac{c_1^2 c_2}{c_0^3} + \frac{40}{9}\frac{c_1 c_3}{c_0^2} + \frac{20}{9}\frac{c_2^2}{c_0^2} - \frac{5}{3}\frac{c_4}{c_0}\right)\right]\\
d_5 &= \frac{1}{3} c_0^{-6/3}\left[\frac{b_5}{5!} - 2\frac{c_1}{c_0}\frac{b_4}{4!} + \frac{b_3}{3!}\left(-2\frac{c_2}{c_0} + 3\frac{c_1^2}{c_0^2}\right) + \frac{b_2}{2!}\left(-2\frac{c_3}{c_0} + 6\frac{c_1 c_2}{c_0^2} - 4\frac{c_1^3}{c_0^3}\right) + \right.\\
&\quad + b_1\left(-2\frac{c_4}{c_0} + 3\frac{c_2^2}{c_0^2} + 6\frac{c_1 c_3}{c_0^2} - 12\frac{c_1^2 c_2}{c_0^3} + 5\frac{c_1^4}{c_0^4}\right) + b_0\left(-2\frac{c_5}{c_0} + 6\frac{c_1 c_4}{c_0^2} + 6\frac{c_2 c_3}{c_0^2} - 12\frac{c_1 c_2^2}{c_0^3} - 12\frac{c_1^2 c_3}{c_0^3} + \right.\\
&\quad \left.\left. + 20\frac{c_1^3 c_2}{c_0^4} - 6\frac{c_1^5}{c_0^5}\right)\right]\\
d_6 &= \frac{1}{3} c_0^{-7/3}\left[\frac{b_6}{6!} - \frac{7}{3}\frac{c_1}{c_0}\frac{b_5}{5!} + \frac{b_4}{4!}\left(-\frac{7}{3}\frac{c_2}{c_0} + \frac{35}{9}\frac{c_1^2}{c_0^2}\right) + \frac{b_3}{3!}\left(-\frac{7}{3}\frac{c_3}{c_0} + \frac{70}{9}\frac{c_1 c_2}{c_0} - \frac{455}{81}\frac{c_1^3}{c_0^3}\right) + \right.\\
&\quad + \frac{b_2}{2!}\left(-\frac{7}{3}\frac{c_4}{c_0} + \frac{35}{9}\frac{c_2^2}{c_0^2} + \frac{70}{9}\frac{c_1 c_3}{c_0^2} - \frac{455}{27}\frac{c_1^2 c_2}{c_0^3} + \frac{1820}{243}\frac{c_1^4}{c_0^4}\right) +\\
&\quad + b_1\left(-\frac{7}{3}\frac{c_5}{c_0} + \frac{70}{9}\frac{c_1 c_4}{c_0^2} + \frac{70}{9}\frac{c_2 c_3}{c_0^2} - \frac{455}{27}\frac{c_1 c_2^2}{c_0^3} - \frac{455}{27}\frac{c_1^2 c_3}{c_0^3} + \frac{7280}{243}\frac{c_1^3 c_2}{c_0^4} - \frac{6916}{729}\frac{c_1^5}{c_0^5}\right) +\\
&\quad + b_0\left(-\frac{7}{3}\frac{c_6}{c_0} + \frac{35}{9}\frac{c_3^2}{c_0^2} + \frac{70}{9}\frac{c_1 c_5}{c_0^2} + \frac{70}{9}\frac{c_2 c_4}{c_0^2} - \frac{455}{81}\frac{c_2^3}{c_0^3} - \frac{455}{27}\frac{c_1^2 c_4}{c_0^3} - \frac{910}{27}\frac{c_1 c_2 c_3}{c_0^3} + \right.\\
&\quad \left.\left. + \frac{7280}{243}\frac{c_1^3 c_3}{c_0^4} + \frac{10920}{243}\frac{c_1^2 c_2^2}{c_0^4} - \frac{34580}{729}\frac{c_1^4 c_2}{c_0^5} + \frac{76076}{6561}\frac{c_1^6}{c_0^6}\right)\right].
\end{aligned} \tag{43}$$

Or, in general,

$$d_n = \frac{1}{3} c_0^{-\frac{n+1}{3}} \sum_{k=0}^{n} \frac{b_{n-k}}{(n-k)!} \sum_{j=0}^{k} \binom{-\frac{n+1}{3}}{j} D_{kj}, \qquad n = 0, 1, 2, \dots,$$

$D_0 = \{C_0\}$

$D_1 = \{0;\ C_1;\}$

$D_2 = \{0;\ C_2;\ C_1^2;\}$

$D_3 = \{0;\ C_3;\ 2\,C_1 C_2;\ C_1^3;\}$

$D_4 = \{0;\ C_4;\ C_2^2 + 2\,C_1 C_3;\ 3\,C_1^2 C_2;\ C_1^4;\}$

$D_5 = \{0;\ C_5;\ 2[C_1 C_4 + C_2 C_3];\ 3[C_1^2 C_3 + C_1 C_2^2];\ 4\,C_1^3 C_2;\ C_1^5\}$

$D_6 = \{0;\ C_6;\ 2[C_1 C_5 + C_2 C_4] + C_3^2;\ C_2^3 + 3\,C_1^2 C_4 + 6\,C_1 C_2 C_3;\ 4\,C_1^3 C_3 + 6\,C_1^2 C_2^2;\ 5\,C_1^4 C_2;\ C_1^6\};$

$$\begin{aligned}
D_7 = \{&0;\ C_7;\ 2[C_1\,C_6 + C_2\,C_5 + C_3\,C_4];\ 3[C_1^2\,C_5 + C_1\,C_3^2 + C_2^2\,C_3 + 2\,C_1\,C_2\,C_4];\\
&4[C_1^3\,C_4 + C_1\,C_2^3];\ 5[C_1^4\,C_3 + 2\,C_1^3\,C_2^2];\ 6\,C_1^5\,C_2;\ C_1^7;\}\\
D_8 = \{&0;\ C_8;\ 2[C_1\,C_7 + C_2\,C_6 + C_3\,C_5] + C_4^2;\ C_1^2\,C_6 + C_2^2\,C_4 + C_2\,C_3^2 + 2\,C_1[C_2\,C_5 + C_3\,C_4];\ 4\,C_1^3\,C_5 + 6\,C_1^2\,C_3^2 +\\
&+ C_2^4 + 12\,C_1\,C_2\,[C_1\,C_4 + C_2\,C_3];\ 5\,[C_1^4\,C_4 + 2\,C_1^2\,C_2^3 + 4\,C_1^3\,C_2\,C_3];\ 3\,C_1^4\,[2\,C_1\,C_3 + 5\,C_2^2];\ 7\,C_1^6\,C_2;\ C_1^8;\}\\
D_9 = \{&0;\ C_9;\ 2[C_1\,C_8 + C_2\,C_7 + C_3\,C_6 + C_4\,C_5];\ C_3^3 + 3[C_1^2\,C_7 + C_2^2\,C_5 + C_1\,C_4^2] +\\
&+ 6[C_1\,C_2\,C_6 + C_1\,C_3\,C_5 + C_2\,C_3\,C_4];\ 4[C_1^3\,C_6 + C_2^3\,C_3 + 3\,C_1\,(C_1\,C_3\,C_4 + C_2^2\,C_4 + C_2\,C_3^2 + C_1\,C_2\,C_5)];\\
&5[C_1^4\,C_5 + C_1\,C_2^4 + 2\,C_1^3\,C_3^2 + 4\,C_1^3\,C_2\,C_4 + 6\,C_1^2\,C_2^2\,C_3];\ 2\,C_1^3\,[3\,C_1^2\,C_4 + 15\,C_1\,C_2\,C_3 + 10\,C_2^3];\\
&7\,C_1^5\,[C_1\,C_3 + 3\,C_2^2];\ 8\,C_1^7\,C_2;\ C_1^9;\}
\end{aligned}$$

$$C_k = c_k/c_0 \tag{43$'$}$$

$D_{kj} = D_{kj}(c_k)$ is a lower triangular matrix the kj-elements of which are given in the row vectors D_k (up to the diagonal element kk), $k = 0, 1, 2, \ldots$.

Hence,

$$\begin{aligned}
d_0 &= \frac{1}{3}\left(\frac{6}{a}\right)^{1/3} a\\
d_1 &= \frac{5}{18}\gamma\left(\frac{6}{a}\right)^{2/3}\\
d_2 &= -\frac{3}{2}\left(\frac{\gamma^2}{4\,a^2} + \frac{3}{5}\beta\right)\\
d_3 &= \frac{1}{9}\left(\frac{6}{a}\right)^{1/3}\left(\frac{a - 28\,\lambda\,a + 28\,\xi\,a'}{5} + \frac{91}{72}\frac{\gamma^3}{a^3} + \frac{7}{10}\beta\,\frac{\gamma}{a}\right)\\
d_4 &= \frac{1}{2}\left(\frac{6}{a}\right)^{2/3}\left\{\frac{[41 - 68\,\lambda - 112\,\xi\,(a'/a)]\,\gamma + 120\,\gamma'\,\xi}{189} - \frac{935}{7776}\frac{\gamma^4}{a^4} + \frac{\beta^2}{70} - \frac{11}{108}\beta\,\frac{\gamma^2}{a^2}\right\}\\
d_5 &= \frac{6}{a^2}\left[-\frac{17}{560}\gamma^2 + \frac{3}{280}\lambda\,\gamma^2 - \frac{93}{1400}\beta\,a^2 + \frac{99}{1400}\lambda\,\beta\,a^2 + \frac{9}{2800}\beta^2\,\gamma - \frac{27}{280}\beta'\,\xi\,a^2 + \frac{9}{350}\beta\,\xi\,a'\,a - \frac{3}{28}\xi\,\gamma'\,\gamma +\right.\\
&\qquad\left. + \frac{1}{32}\beta\,\frac{\gamma^3}{a^2} + \frac{3}{20}\gamma^2\,\frac{\xi\,a'}{a} + \frac{7}{256}\frac{\gamma^5}{a^4}\right]\\
d_6 &= \left(\frac{6}{a}\right)^{1/3}\left[-\frac{133}{540}\lambda\,a + \frac{98}{540}\lambda^2\,a + \frac{1}{16}\xi\,\beta'\,\frac{\gamma}{a} + \frac{1}{12}\xi\,\gamma'\,\frac{\beta}{a} - \frac{7}{27}\xi\,\lambda'\,a + \frac{77}{135}\xi\,a' - \frac{63}{270}\xi\,a'\,\lambda + \frac{7}{27}\xi^2\,a'' -\right.\\
&\qquad - \frac{28}{135}\frac{(\xi\,a')^2}{a} - \frac{1}{540}a + \frac{65}{216}\xi\,\gamma'\,\frac{\gamma^2}{a^3} + \frac{1}{300}\frac{\beta^3}{a} + \frac{67}{2160}\beta\,\frac{\gamma}{a} + \frac{23}{2160}\lambda\,\beta\,\frac{\gamma}{a} + \frac{143}{1944}\frac{\gamma^3}{a^3} + \frac{65}{3888}\lambda\,\frac{\gamma^3}{a^3} -\\
&\qquad\left. - \frac{107}{540}\xi\,a'\,\beta\,\frac{\gamma}{a^2} - \frac{13}{432}\beta^2\,\frac{\gamma^2}{a^3} - \frac{1729}{15552}\beta\,\frac{\gamma^4}{a^5} - \frac{43225}{559872}\frac{\gamma^6}{a^7} - \frac{455}{972}\xi\,a'\,\frac{\gamma^3}{a^4}\right];
\end{aligned} \tag{44}$$

by Eq. (43) upon elimination of the b_n and c_n given in Eqs. (35) and (37). For practical applications, the expansion parameter $a(\xi)$ is evaluated from Eqs. (42) and (44) by successive approximation or numerical iteration. To this end, Eq. (42) is rearranged, if necessary as a convergent sum by means of EULER's transformation [7]

$$\sum_{m=0}^{\infty} A_m = \sum_{m=0}^{\infty} \frac{B_m}{2^{m+1}}, \qquad B_m = \binom{m}{0} A_0 + \binom{m}{1} A_1 + \binom{m}{2} A_2 + \ldots \binom{m}{m} A_m .$$

Separated Flow

At the point of separation, the normal derivative of the longitudinal velocity component vanishes at the plate, $[\partial u/\partial\eta]_{\eta=0} = 0$, i. e. by Eqs. (15) and (28)

$$a(\xi) = a_2(\xi) = [\partial^2 f/\partial\eta^2]_{\eta=0} = 0 . \tag{45}$$

The expansion for $f(\xi, \eta)$ [Eq. (28)] starts with η^3 at the separation point,

$$f(\xi, \eta) = \sum_{n=3}^{\infty} a_n \frac{\eta^n}{n!} \tag{46}$$

where

$$\begin{aligned}
a_3 &= \gamma, \qquad a_5 = -\beta\,\gamma,\\
a_7 &= -(4 + 6\,\lambda)\,\gamma^2 + 4\,\xi\,\gamma'\,\gamma + \beta^2\,\gamma\\
a_9 &= \beta\,[6(5 + 6\,\lambda)\,\gamma^2 - \gamma\,\beta^2] - 6\,\xi\,(2\,\beta\,\gamma' + 3\,\beta'\,\gamma)\,\gamma + 36\,\gamma(\xi\,a')^2\\
a_{11} &= \beta^4\,\gamma - 6\,(25 + 27\,\lambda)\,\beta^2\,\gamma^2 + 16\,(2 + 3\,\lambda)\,(8 + 7\,\lambda)\,\gamma^3 - 16\,(12 + 23\,\lambda)\,\xi\,\gamma'\,\gamma^2\\
&\quad - 240\,\xi\,\lambda'\,\gamma^3 + 24\,\xi\,\gamma'\,\beta^2\,\gamma + 126\,\xi\,\beta'\,\beta\,\gamma^2 - 64\,(\xi\,\gamma')^2\,\gamma + 160\,\xi^2\,\gamma''\,\gamma^2 - 36\,(\xi\,a')^2\,\beta\,\gamma
\end{aligned}$$

$$a_{2n} = 0 \tag{47}$$

since for separation $a(\xi) = 0$. The corresponding expansion for $F(\xi, \eta)$ [Eq. (29)] starts with η^4,

$$F(\xi, \eta) = \eta^4 \sum_{n=0}^{\infty} c_n\,\eta^n = \tau, \qquad c_n = \frac{a_{n+3}}{(n + 4)!} . \tag{48}$$

The point of separation $\xi_s = \xi$ is determined by the boundary condition in Eq. (20), which becomes in view of Eq. (25)

$$f'(\xi,\eta)_{\eta=\infty} = \int\limits_0^\infty e^{-F(\xi,\eta)}\,\varphi(\xi,\eta)\,d\eta = 1 \tag{49}$$

where

$$\varphi(\xi,\eta) = \sum_{n=1}^{\infty} b_n \frac{\eta^n}{n!} \tag{50}$$

and

$$\begin{aligned}
&b_1 = \gamma\,, \qquad b_3 = -\beta\gamma\\
&b_5 = (1-6\lambda)\gamma^2 + \beta^2\gamma + 4\,\xi\gamma'\gamma\\
&b_7 = \beta[12\,(3\lambda-1)\gamma - \beta^2]\,\beta\gamma - 6\,\xi\,(2\,\beta\gamma' + 3\,\beta'\gamma)\gamma + 36\,(\xi\,a')^2\gamma\\
&b_9 = \gamma\{2(2+3\lambda)(56\lambda-71)\gamma^2 + 6\,(64-27\lambda)\gamma\beta^2 + \beta^4 + 4\,(87-92\lambda)\,\xi\gamma'\gamma -\\
&\quad - 240\,\xi\lambda'\gamma^2 + 24\,\xi\gamma'\beta^2 + 12\,6\,\xi\beta'\beta\gamma - 36\,\beta(\xi\,a')^2 - 64(\xi\gamma')^2 + 160\,\xi\gamma''\gamma\}\\
&\overline{}\\
&b_{2n} = 0
\end{aligned} \tag{51}$$

since for separation $a(\xi) = 0$. For the evaluation of the integral in Eq. (49) let

$$\varphi(\xi,\eta)\frac{d\eta}{d\tau} = \sum_{n=0}^{\infty} d_n\,\tau^{(n-3)/4} \tag{52}$$

where the expansion starts with $\tau^{-3/4}$, since η starts with $\tau^{1/4}$ [Eq. (48)]. Accordingly [7],

$$\int\limits^{(+0)} \frac{\varphi(\xi,\eta)\,d\eta}{\tau^{(n+1)/4}} = d_n \int\limits^{(+0)(+0)(+0)(+0)} \frac{d\tau}{\tau} = 8\,\pi\,i\,d_n\,,$$

i. e.

$$d_n = \frac{1}{8\,\pi\,i}\int\limits^{(+0)} \frac{\varphi(\xi,\eta)\,d\eta}{\tau^{(n+1)/4}}\,. \tag{53}$$

Eq. (53) indicates that d_n is one-fourth of the coefficient of η^n in the expansion of the right side of the equation

$$\varphi(\xi,\eta)\,\tau^{-(n+1)/4} = \left(\frac{b_1}{1!}\eta^1 + \frac{b_3}{3!}n^3 + \frac{b_5}{5!}\eta^5 + \ldots\right)(c_0 + c_2\eta^2 + c_3\eta^3 + \ldots)^{-(n+1)/4}\,. \tag{54}$$

Thus,

$$\begin{aligned}
&d_1 = \frac{1}{4}c_0^{-1/2}b_1\\
&d_3 = \frac{1}{4}c_0^{-1}\left[\frac{b_3}{3!} - \frac{c_2}{c_0}b_1\right]\\
&d_5 = \frac{1}{4}c_0^{-3/2}\left[\frac{b_5}{5!} - \frac{3}{2}\frac{c_2}{c_0}\frac{b_3}{3!} + \left(\frac{15}{8}\frac{c_2^2}{c_0^2} - \frac{3}{2}\frac{c_4}{c_0}\right)b_1\right]\\
&d_7 = \frac{1}{4}c_0^{-4/2}\left[\frac{b_7}{7!} - 2\frac{c_2}{c_0}\frac{b_5}{5!} + \left(3\frac{c_2^2}{c_0^2} - 2\frac{c_4}{c_0}\right)\frac{b_3}{3!} + \left(-2\frac{c_6}{c_0} + 6\frac{c_2c_4}{c_0^2} - 4\frac{c_2^3}{c_0^3}\right)b_1\right]\\
&d_9 = \frac{1}{4}c_0^{-5/2}\left[\frac{b_9}{9!} - \frac{5}{2}\frac{c_2}{c_0}\frac{b_7}{7!} + \left(\frac{35}{8}\frac{c_2^2}{c_0^2} - \frac{5}{2}\frac{c_4}{c_0}\right)\frac{b_5}{5!} + \left(-\frac{5}{2}\frac{c_6}{c_0} + \frac{35}{4}\frac{c_2c_4}{c_0^2} - \frac{105}{16}\frac{c_2^3}{c_0^3}\right)\frac{b_3}{3!} +\right.\\
&\quad\left. + \left(-\frac{5}{2}\frac{c_8}{c_0} + \frac{35}{8}\frac{c_4^2}{c_0^2} + \frac{35}{4}\frac{c_2c_6}{c_0^2} - \frac{315}{16}\frac{c_2^2c_4}{c_0^3} + \frac{1155}{128}\frac{c_2^4}{c_0^4}\right)b_1\right]\\
&\overline{}\\
&d_{2n} = 0;
\end{aligned} \tag{55}$$

hence

$$\begin{aligned}
&d_1 = \sqrt{\frac{3}{2}}\,\gamma^{1/2}\,, \qquad d_3 = -\frac{4}{5}\beta\\
&d_5 = \sqrt{\frac{3}{2}}\,\gamma^{-1/2}\left[\frac{1}{14}(4-15\lambda)\gamma + \frac{5}{7}\xi\gamma' + \frac{1}{35}\beta^2\right]\\
&d_7 = \frac{4}{25}\left[\frac{2}{7}(9\lambda-8)\beta + \frac{4}{25}\frac{\beta^3}{\gamma} + \frac{2}{7}\frac{\beta}{\gamma}(\xi\gamma') - 3\,\xi\beta' + \frac{45}{7}\frac{1}{\gamma}(\xi\,a')^2\right]\\
&d_9 = \frac{1}{2310}\sqrt{\frac{3}{2}}\,\gamma^{-1/2}\left[-985\,\gamma - 644\,\lambda\gamma + \frac{6021}{5}\beta^2 - \frac{63}{5}\lambda\beta^2 - \frac{63}{5}\lambda\beta^2 + \frac{2625}{4}\gamma\lambda^2 + \frac{49}{25}\frac{1}{\gamma}\beta^4 +\right.\\
&\quad + 1506\,\xi\gamma' - 840\,\gamma(\xi\lambda') + \frac{122}{5}\beta(\xi\beta') - 595\,\lambda(\xi\gamma') - \frac{42}{5}\frac{\beta^2}{\gamma}(\xi\gamma') + 660\,\frac{\beta}{\gamma}(\xi\,a')^2 - 455\,\frac{1}{\gamma}(\xi\gamma')^2 +\\
&\quad\left. + 560\,\xi^2\gamma''\right]\\
&\overline{}\\
&d_{2n} = 0
\end{aligned} \tag{56}$$

after elimination of b_n and c_n by means of Eqs. (48) and (51). The condition for separation in Eq. (49) becomes upon substitution of Eqs. (48) and (52)

$$\sum_{n=1}^{\infty} d_n \Gamma\left(\frac{n+1}{4}\right) = 1, \qquad d_n = d_n(\xi). \tag{57}$$

This is an equation in ξ, which determines the point of separation $\xi_s = \xi$. Since the terms with increasing n contribute less and less to the sum, ξ is readily evaluable by successive approximation.

Application

As a concrete illustration, the above formalism is used for the determination of the boundary-layer separation in a conducting diffuser flow with a transverse, inhomogeneous magnetic field, $B_y(x, y)$. The free stream velocity of a diffuser is [10].

$$u_\infty(\bar{x}) = u_0/\bar{x}, \qquad 1 \leq \bar{x} < \infty. \tag{58}$$

The associated nondimensional, spatial coordinates are

$$\bar{x} = x/d, \tag{59}$$

$$\bar{y} = \sqrt{R/2}\,(y/d). \tag{60}$$

The REYNOLDS (R) and HARTMANN (M) numbers of the diffuser are defined as

$$R = u_0 d/\nu, \qquad M^2 = (\sigma/\varrho\,\nu)\, B_0^2 d^2 \tag{61}$$

where B_0 is a magnetic field constant associated with the field $B_y(x, y)$. By Eqs. (21), (22), and (32),

$$\lambda(\bar{x}) = 2 \ln \bar{x}, \tag{62}$$

$$\beta(\bar{x}) = -2\frac{M^2}{R}\left(\frac{B_y}{B_0}\right)^2 \bar{x}^2 \ln \bar{x}, \tag{63}$$

$$\gamma(\bar{x}) = 2\left[1 - \frac{M^2}{R}\left(\frac{B_y}{B_0}\right)^2 \bar{x}^2\right] \ln \bar{x}. \tag{64}$$

Hence,

$$\xi\,\lambda' = \bar{x} \ln \bar{x} \frac{d\lambda}{d\bar{x}} = 2 \ln \bar{x} = \lambda, \tag{65}$$

$$\xi\,\beta' = \bar{x} \ln \bar{x} \frac{d\beta}{d\bar{x}} = -2\frac{M^2}{R}\left(\frac{B_y}{B_0}\right)^2 \left\{1 + 2 \ln \bar{x}\left[1 + \frac{d \ln B_y}{d \ln \bar{x}}\right]\right\} \bar{x}^2 \ln \bar{x}, \tag{66}$$

$$\xi\,\gamma' = \bar{x} \ln \bar{x} \frac{d\gamma}{d\bar{x}} = 2\left[1 - \frac{M^2}{R}\left(\frac{B_y}{B_0}\right)^2 \left\{1 + 2 \ln \bar{x}\left[1 + \frac{d \ln B_y}{d \ln \bar{x}}\right]\right\} \bar{x}^2\right] \ln \bar{x}, \tag{67}$$

and

$$\xi\, a' = \bar{x} \ln \bar{x} \frac{da}{d\bar{x}}, \qquad \xi^2 a'' = \bar{x} (\ln \bar{x})^2 \left[\frac{da}{d\bar{x}} + \bar{x}\frac{d^2 a}{d\bar{x}^2}\right], \quad \text{etc.}, \tag{68}$$

since

$$\xi = \ln \bar{x}, \quad \eta = \bar{y}/(\bar{x}\sqrt{\ln \bar{x}}) \tag{69}$$

by Eqs. (14) and (60). The dimensionless velocity components $\bar{u}$ and $\bar{v}$ in the diffuser boundary-layer are by Eqs. (15) and (16)

$$\bar{u} \equiv \frac{u}{u_0} = \frac{1}{\bar{x}} \frac{\partial f}{\partial \eta}, \tag{70}$$

$$\bar{v} \equiv \sqrt{\frac{R}{2}}\frac{v}{u_0} = -\frac{1}{2\bar{x}\sqrt{\ln \bar{x}}}\left[f + 2\xi \frac{\partial f}{\partial \xi} - (1 + 2 \ln \bar{x})\, \eta \frac{\partial f}{\partial \eta}\right], \tag{71}$$

where

$$f = \sum_{n=2}^{\infty} \frac{a_n(\bar{x})}{n!}\left(\frac{\bar{y}}{\bar{x}\sqrt{\ln \bar{x}}}\right)^n, \tag{72}$$

$$\xi \frac{\partial f}{\partial \xi} = \bar{x} \ln \bar{x} \sum_{n=2}^{\infty} \frac{da_n(\bar{x})}{n!\, d\bar{x}}\left(\frac{\bar{y}}{\bar{x}\sqrt{\ln \bar{x}}}\right)^n, \tag{73}$$

$$\eta \frac{\partial f}{\partial \eta} = \sum_{n=2}^{\infty} \frac{a_n(\bar{x})}{(n-1)!}\left(\frac{\bar{y}}{\bar{x}\sqrt{\ln \bar{x}}}\right)^n. \tag{74}$$

The separation condition in Eq. (57) becomes upon substitution of the $d_n(\bar{x})$ from Eq. (56)

1. *Approximation,* $n \leqq 1$:

$$\gamma^{1/2} = \sqrt{\frac{2}{3\pi}}\,; \tag{75}$$

2. *Approximation,* $n \leqq 3$:

$$\gamma^{1/2} = \sqrt{\frac{2}{3\pi}}\left(1 + \frac{4}{5}\beta\right); \tag{76}$$

3. *Approximation,* $n \leqq 5$:

$$\gamma^{1/2} = \sqrt{\frac{2}{3\pi}}\left\{1 + \frac{4}{5}\beta - \frac{1}{2}\sqrt{\frac{3\pi}{2}}\gamma^{1/2}\left[\frac{2}{7} - \frac{15}{14}(\gamma - \beta) + \frac{5}{7}\frac{(\xi\gamma')}{\gamma} + \frac{1}{35}\frac{\beta^2}{\gamma}\right]\right\}; \tag{77}$$

4. *Approximation,* $n \leqq 7$:

$$\gamma^{1/2} = \sqrt{\frac{2}{3\pi}}\left\{1 + \frac{4}{5}\beta - \frac{1}{2}\sqrt{\frac{3\pi}{2}}\gamma^{1/2}\left[\frac{2}{7} - \frac{15}{14}(\gamma - \beta) + \frac{5}{7}\frac{(\xi\gamma')}{\gamma} + \frac{1}{35}\frac{\beta^2}{\gamma}\right] - \right.$$
$$\left. - \frac{4}{25}\left[\frac{18}{7}(\gamma - \beta)\beta - \frac{16}{7}\beta + \frac{4}{25}\frac{\beta^3}{\gamma} + \frac{2}{7}\frac{\beta}{\gamma}(\xi\gamma') - 3(\xi\beta') + \frac{45}{7\gamma}(\xi a')^2\right]\right\}; \tag{78}$$

5. *Approximation,* $n \leqq 9$:

$$\gamma^{1/2} = \sqrt{\frac{2}{3\pi}}\left\{1 + \frac{4}{5}\beta - \frac{1}{2}\sqrt{\frac{3\pi}{2}}\gamma^{1/2}\left[\frac{2}{7} - \frac{15}{14}(\gamma - \beta) + \frac{5}{7}\frac{(\xi\gamma')}{\gamma} + \frac{1}{35}\frac{\beta^2}{\gamma}\right] - \frac{4}{25}\left[\frac{18}{7}(\gamma - \beta)\beta - \right.\right.$$
$$- \frac{16}{7}\beta + \frac{4}{25}\frac{\beta^3}{\gamma} + \frac{2}{7}\frac{\beta}{\gamma}(\xi\gamma') - 3(\xi\beta') + \left.\frac{45}{7\gamma}(\xi a')^2\right] - \frac{(3/4)}{2310}\sqrt{\frac{3\pi}{2}}\gamma^{1/2}\left[-985 - 1484(\gamma - \beta) + \right.$$
$$+ \frac{6021}{5}\frac{\beta^2}{\gamma} - \frac{63}{5}(\gamma - \beta)\frac{\beta^2}{\gamma} + \frac{2625}{4}(\gamma - \beta)^2 + \frac{49}{25}\frac{\beta^4}{\gamma^2} + 1506\frac{(\xi\gamma')}{\gamma} + \frac{122}{5}\frac{\beta}{\gamma}(\xi\beta') -$$
$$\left.\left. - 595(\gamma - \beta)\frac{(\xi\gamma')}{\gamma} - \frac{42}{5}\frac{\beta^2}{\gamma^2}(\xi\gamma') + 660\frac{\beta}{\gamma^2}(\xi a')^2 - 455\frac{(\xi\gamma')^2}{\gamma^2} + 560\frac{\xi^2\gamma''}{\gamma}\right]\right\}. \tag{79}$$

In the vicinity of the origin of the diffuser, i. e. the possible separation region $\bar{x} \gtrsim 1$, an inhomogeneous (growing) magnetic field may be simulated analytically by a linear dependence [Eq. (3)] on the spatial coordinates of the form

$$B_y = B_0\left(1 + \frac{x}{l}\right) = B_0(1 + \varepsilon\bar{x}) \qquad \varepsilon = \frac{d}{l} \geqq 0\,,$$
$$B_x = B_1 - B_0\frac{y}{l} = B_1\left(1 - \varepsilon\frac{B_0}{B_1}\sqrt{\frac{2}{R}}\bar{y}\right) \tag{80}$$

where the dimensionless quantity ε characterizes the inhomogeneity of B_y. The transverse component has the properties

$$\frac{d\ln B_y}{d\ln\bar{x}} \equiv \frac{\varepsilon\bar{x}}{1 + \varepsilon\bar{x}} = \begin{cases}0, & \varepsilon \to 0;\\ 1, & \varepsilon \gg 1, \quad \bar{x} \gtrsim 1\,.\end{cases} \tag{81}$$

With regard to experimental applications, one is primarily interested I) in magnetic fields which are noticeably inhomogeneous in the entrance region ($\bar{x} \gtrsim 1$) and II) in large REYNOLDS number flows for which

$$\text{I)}\ \left(\frac{B_y}{B_0}\right)^2 \equiv (1 + \varepsilon\bar{x})^2 > 1\,, \qquad \text{II)}\ \frac{M^2}{R} \ll 1\,.$$

In the derivation of an approximate separation criterion, it will, therefore, be assumed that

$$\frac{M^2}{R}\left(\frac{B_y}{B_0}\right)^2 \lesssim 1\,. \tag{82}$$

The first approximation, $\gamma(\bar{x}) = 2/(3\pi)$ [Eq. (75)], indicates then that $M^2/R \lesssim 0.35$ and $\bar{x} \lesssim 1.3$ for separation ($M^2/R \approx 0.35$ for $\varepsilon = 0$). Accordingly, if Eq. (82) is satisfied,

$$0 \leqq |\beta(\bar{x})| \equiv 2\frac{M^2}{R}\left(\frac{B_y}{B_0}\right)^2\bar{x}^2\ln\bar{x} \lesssim 0.3,\ \varepsilon \geqq 0,\ 1 \leqq \bar{x} \lesssim 1.3\,. \tag{83}$$

According to Eqs. (63)—(67) and (81), the derivative terms show the asymptotic behavior ($\sim$ for proportional)

$$\beta' \xi \equiv \beta + (\gamma - \beta)\beta\left(1 + \frac{d \ln B_y}{d \ln \bar{x}}\right) \sim \beta, \qquad |\beta| \ll 1;$$

$$\xi \gamma' \equiv \gamma + (\gamma - \beta)\beta\left(1 + \frac{d \ln B_y}{d \ln \bar{x}}\right) \cong \gamma, \qquad |\beta| \ll 1; \tag{84}$$

$$\xi^2 \gamma'' \equiv \xi \frac{d}{d\xi}\left[(\gamma - \beta)\beta\left(1 + \frac{d \ln B_y}{d \ln \bar{x}}\right)\right] \sim \beta, \qquad |\beta| \ll 1.$$

A separation criterion, which holds for the subcase $|\beta| \ll 1$, will be in good approximation generally valid, since the maximum possible value of $|\beta|$ is already significantly smaller than 1 in the case of separation, i. e. $|\beta|_{max} \cong 0.3$ by Eq. (83). The Eqs. (74)—(79) give, in the sense of a successive approximation, the following separation criteria for $|\beta| \ll 1$:

$$\gamma_1^{1/2} = \sqrt{\frac{2}{3\pi}} = \gamma_2^{1/2}; \tag{85}$$

$$\gamma_3^{1/2} = \sqrt{\frac{2}{3\pi}}\left[1 - \frac{1}{2}\left(1 - \frac{5}{7\pi}\right)\right] = \gamma_4^{1/2}; \tag{86}$$

$$\gamma_5^{1/2} = \sqrt{\frac{2}{3\pi}}\left\{1 - \frac{73}{140}\left[1 - \frac{1}{2}\left(1 - \frac{5}{7\pi}\right)\right]^1 + \frac{113}{140\pi}\left[1 - \frac{1}{2}\left(1 - \frac{5}{7\pi}\right)\right]^3 - \frac{175}{1848\pi^2}\left[1 - \frac{1}{2}\left(1 - \frac{5}{7\pi}\right)\right]^5\right\} = \gamma_6^{1/2}. \tag{87}$$

where the right sides of the various approximations n have been evaluated by means of the preceding $(n - 1)$-th approximations, $n = 1, \ldots, 5$. Accordingly,

$$\gamma_{3,4}/\gamma_{1,2} \cong (1 - 0.4)^2, \qquad \gamma_5/\gamma_{1,2} \cong (1 - 0.3)^2$$

or

$$\gamma_n = \gamma_{n-1}(1 + \delta_n)^2, \qquad n \geqq 2$$

$$\delta_2 \cong 0, \quad \delta_3 \cong -0.4, \quad \delta_4 = 0, \quad \delta_5 \cong +0.2,$$

which indicates a tendency of convergence. It follows as separation criterion in the case $|\beta| \ll 1$:

$$\gamma(\bar{x}) \cong 1/(3\pi). \tag{88}$$

This result differs from the first approximation in Eq. (85) by a correction factor $(1 - 0.3)^2 \cong 1/2$. In applying Eq. (88) in the entire range $0 \leqq |\beta| \cong 0.3$ defined in Eq. (83), it should be noted that the accuracy decreases as $|\beta|$ increases towards $|\beta|_{max}$.

Upon substitution for $\gamma(\bar{x})$ and $B_y(\bar{x})$ in accordance with the Eqs. (64) and (80), the separation condition in Eq. (88) becomes:

$$\frac{M^2}{R} = \left(1 - \frac{1}{6\pi \ln \bar{x}}\right)\bar{x}^{-2}(1 + \varepsilon\bar{x})^{-2}. \tag{89}$$

Eq. (89) gives the separation point $x = \bar{x}\, d$ in dependence of the experimental flow parameters M^2 and R. The possible points of separation are restricted to the region

$$\bar{x}_{min} \leqq \bar{x} \leqq \bar{x}_{max} \tag{90}$$

where

$$\bar{x}_{min} = \exp\left(\frac{1}{6\pi}\right) = 1.054 \tag{91}$$

and

$$\bar{x}_{max} = \exp\left\{\frac{1}{12\pi}\left[1 + \left(1 + 12\pi\frac{1 + \varepsilon\bar{x}_{max}}{1 + 2\varepsilon\bar{x}_{max}}\right)^{1/2}\right]\right\}, \tag{92}$$

i. e.

$$\bar{x}_{max} = \exp[(1 + \sqrt{1 + 12\pi})/12\pi] = 1.211, \qquad \text{for } \varepsilon \ll 1, \tag{93}$$

$$\bar{x}_{max} \cong \exp[(1 + \sqrt{1 + 6\pi})/12\pi] = 1.156, \qquad \text{for } \varepsilon \geqq 10, \tag{94}$$

Hence, separation occurs if the flow parameters M and R satisfy the conditions

$$0 \leqq \frac{M^2}{R} \leqq \left(\frac{M^2}{R}\right)_{max} = \left[1 - \frac{1}{6\pi \ln \bar{x}_{max}}\right]\bar{x}_{max}^{-2}(1 + \varepsilon\bar{x}_{max})^{-2}, \tag{95}$$

where

$$(M^2/R)_{\max} = 0.493\,, \qquad \text{for } \varepsilon \ll 1\,, \tag{96}$$

$$(M^2/R)_{\max} \cong 0.356\,\varepsilon^{-2}, \qquad \text{for } \varepsilon \gtrsim 10\,. \tag{97}$$

If the actual M^2/R of the flow is larger than $(M^2/R)_{\max}$, separation is no longer possible. By Eq. (95), a magnetic field which grows in the downstream direction may effectively suppress separation, if the inhomogeneity parameter ε is sufficiently large. Note that the restriction in Eq. (82) imposes an upper limit on the ε-values, for which Eq. (95) represents a valid approximation, $0 \leqq \varepsilon \lesssim 10$.

In Fig. 2, M^2/R versus $\bar{x}$ is represented according to the improved separation criterion in Eq. (89), with ε as a parameter, $\varepsilon = 0$, 10^{-1}, 10^0, and 10^1. The intersection of the separation curve with the abscissa is at $\bar{x}_{\min} = 1.054$ and the abscissa of the maximum of the separation curve is $\bar{x}_{\max}$, which depends slightly on ε. It is seen that the domain of M^2/R within which separation is feasible is drastically reduced with increasing ε.

The determination of M^2/R in dependence of $\bar{x}$ and ε from Eq. (79) requires numerical iteration. By comparison with the exact iterative solution, it can be shown that Eq. (89) gives a good approximation to the separation point $\bar{x}$ in the region $\bar{x}_{\min} < \bar{x} \leqq \bar{x}_{\max}$ and to the critical value $(M^2/R)_{\max}$ for any ε. The "laminar reattachment" region, $\bar{x} > \bar{x}_{\max}$, has no physical meaning (breakdown of laminar boundary-layer approximation after separation).

Experiments with compressible plasma flows[1,11] confirm qualitatively the theoretical results according to which separation can be inhibited by means of strongly inhomogeneous magnetic fields. On the other hand, there seem to be no experimental data yet available on separation in incompressible flows of conducting fluids across inhomogeneous magnetic fields. For the latter reason, a quantitative comparison beetwen theoretical prediction and experiment cannot be given.

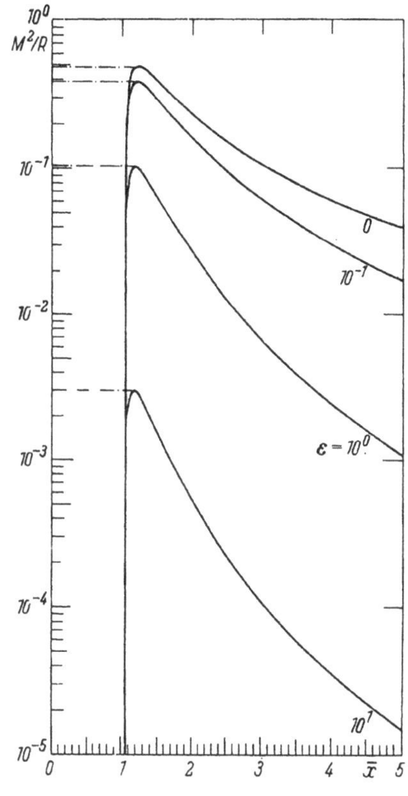

Fig. 2. M^2/R versus $\bar{x}$ for $\varepsilon = 0, 10^{-1}, 10^0$ and 10^1 showing suppression of separation

Note: The coefficients d_n in Eq. (44) reduce for $\beta = 0$ and $\gamma = \lambda$ to the d_n-coefficients of MEKSYN [7, p. 100, Eq. (3.13)] except in the cases d_4 and d_5. The latter coefficients of MEKSYN [7] seem to be in error, see also Liron [12].

References

1 ZAUDERER, B., Phys. Fluids 11, 2577 (1968).
2 KIRKO, J. M., Magnetohydrodynamics of Liquid Metals, Consultants Bureau, New York 1965.
3 See, e. g.: ROSSOW, V. J., ZAMP, **19b**, 519 (1958). LEIBOVICH, S., J. Fluid Mech. **29**, 401 (1967). KATAGARI, M., J. Phys. Soc. Japan **27**, 1045 (1969). **27**, 1051 (1969). **27**, 1662 (1962).
4 MEKSYN, D., Proc. Roy. Soc. A 237, 543 (1956).
5 MEKSYN, D., J. Aero/Space Sci. **29**, 662 (1961).
6 MEKSYN, D., ZAMP, **17**, 397 (1966).
7 MEKSYN, D., Laminar Boundary-Layer Theory, Pergamon Press, New York 1961.
8 The MHD boundary-layer requires electrodes in planes $z = \pm c$ to collect the induced current density, which impress via an external circuit an electric field E_z.
9 ABRAMOWITZ, M., and STEGUN, I. A., Handbook of Mathematical Functions, Dover, New York 1965.
10 POHLHAUSEN, K., ZAMM **1**, 252 (1921).
11 ZAUDERER, B., AIAA J. **7**, 2189 (1969).
12 LIRON, N., ZAMP **24**, 181 (1973).

Eingereicht am 18. 1. 1973

Anschrift: Dr. N. LIRON, The Weizmann Institute of Science, Department of Applied Mathematics, Rehovot, Israel;
Professor H. E. WILHELM, Colorado State University, Department of Physics, Fort Collins, Colorado 80521, USA

ZAMM 54, 39 – 51 (1974)

A. NEUMAIER

Rundungsfehleranalyse einiger Verfahren zur Summation endlicher Summen

Der bei einer Summation auftretende Rundungsfehler kann als Maß für die Güte des verwendeten Verfahrens gelten. Im folgenden werden für mehrere Summierungsverfahren, unter anderem für das übliche und das Kahan-Babuška-Verfahren, a-priori-Schranken für diese Rundungsfehler angegeben und miteinander verglichen.

The rounding-error arising during summation can be interpreted as a measure for the quality of the procedure used. In the following, a-priori-bounds for this rounding-error are used to compare several summation procedures, e.g. the common procedure and the method of Kahan-Babuška.

Возникающие при суммировании ошибки скругления могут рассматриваться как мера качества для применяемого метода. В последующем приведены и сравнены между собой для многих методов суммирования, в том числе для "обычного" метода и для метода Каган-Бабуска, априорные пределы для этих ошибок скругления.

1. Aufgabenstellung

Zur Berechnung einer endlichen Summe

$$s := \sum_{m=1}^{n} a_m \tag{1}$$

können verschiedene Verfahren benutzt werden. Bei Durchführung der Rechnung im Körper der reellen Zahlen liefert jedes Verfahren das Ergebnis s; wird die Rechnung aber auf einer Rechenmaschine durchgeführt, so ergeben sich im allgemeinen verschiedene Ergebnisse, da die auftretenden Rundungsfehler vom Verfahren abhängen.

Die *Güte* einer Summationsvorschrift kann z. B. durch das Verhältnis von Genauigkeit des berechneten Wertes zum Arbeitsaufwand angegeben werden. Gesucht sind daher Schranken für die Abweichung $\tilde{s} - s$ des nach einem bestimmten Verfahren berechneten Näherungswertes $\tilde{s}$ für s.

2. Summationsvorschriften

Es werden die folgenden vier Summationsvorschriften untersucht[1]):

I. Normalverfahren

Man setzt

$$\begin{aligned} s_0 &:= 0 \\ s_m &:= s_{m-1} + a_m \qquad \text{für} \quad m = 1, \ldots, n \\ s &:= s_n , \end{aligned} \tag{I,1}$$

d. h., die Summe (1) wird von links nach rechts abgearbeitet, z. B.

$$s = \big((a_1 + a_2) + a_3\big) + a_4 \qquad (n = 4)$$

Der Rechenaufwand beträgt $n - 1$ Additionen (Addition von 0 wird vernachlässigt).

II. Verbessertes Normalverfahren

2^t sei die kleinste Potenz von 2 mit $2^t \geqq n$. Setzt man

$$\begin{aligned} s_{m0} &:= a_m \qquad \text{für} \quad m = 1, \ldots, n \\ s_{m0} &:= 0 \qquad \text{für} \quad m = n + 1, \ldots, 2^t , \end{aligned}$$

so berechnet auch die Vorschrift

$$\begin{aligned} s_{mk} &:= s_{2m-1,k-1} + s_{2m,k-1} \qquad \text{für} \quad 1 \leqq k \leqq t , \quad 0 \leqq m \leqq 2^{t-k} - 1 \\ s &:= s_{1t} \end{aligned} \tag{II,1}$$

die Summe s, aber in anderer Klammerung, z. B.

$$s = \big((a_1 + a_2) + (a_3 + a_4)\big) + \big((a_5 + a_6) + (a_7 + a_8)\big) \qquad (t = 3,\ n = 8) .$$

Wieder sind $n - 1$ Additionen auszuführen.

[1]) BABUŠKA [1] behandelte die Verfahren I, II, III unter den Bezeichnungen L_{1A}, L_{1B}, L_{1C}.

III. KAHAN-BABUŠKA-Verfahren

Man setzt [1], [2], [3]

$$s_0 := 0\,, \qquad s_m := a_m + s_{m-1} \qquad (m = 1, \ldots, n)$$
$$w_0 := 0\,, \qquad w_m := w_{m-1} + \big(a_m + (s_{m-1} - s_m)\big) \qquad (m = 1, \ldots, n) \qquad \text{(III,1)}$$
$$s := s_n + w_n\,.$$

Die w_m sind bei reeller Rechnung Null; wird auf einer Maschine gerechnet, so sammeln sich die Rundungsfehler in diesem Term an. Durch die Korrektur $+ w_n$ werden die Rundungsfehler etwa kompensiert. Es ist also eine größere Genauigkeit zu erwarten. Der Rechenaufwand beträgt hier $4n - 3$ Additionen/Subtraktionen, ist also etwa viermal so groß wie in Verfahren I und II.

IV. Verbessertes KAHAN-BABUŠKA-Verfahren

Man setzt

$$s_0 := 0\,, \qquad s_m := a_m + s_{m-1} \qquad (m = 1, \ldots, n)$$
$$w_0 := 0\,, \qquad w_m := w_{m-1} + \big(a_m + (s_{m-1} - s_m)\big)\,, \quad \text{wenn } |a_m| \leqq |s_{m-1}| \qquad \text{(IV,1)}$$
$$w_m := w_{m-1} + \big(s_{m-1} + (a_m - s_m)\big)\,, \quad \text{wenn } |a_m| > |s_{m-1}|$$
$$s := s_n + w_n\,.$$

Das Verfahren unterscheidet sich von Verfahren III nur für $|a_m| > |s_{m-1}|$. (Z. B. ist IV = III, wenn (a_m) eine monoton fallende Folge positiver reeller Zahlen ist, da $|a_m| > |s_{m-1}|$ genau für $m = 1$.) Hier sind außer $4n - 3$ Additionen/Subtraktionen noch n Abfragen auszuführen. Rechnet man für eine Abfrage denselben Aufwand wie für eine Subtraktion, so ist der Aufwand für Verfahren IV etwa fünfmal so groß wie für Verfahren I oder II, und 1/4mal größer als in Verfahren III.

3. Rundungsfehler beim Normalverfahren (I und II)

Nach WILKINSON [4] genügt die Maschinennäherung $a \tilde{+} b$ einer Summe $a + b$ für viele Rechenmaschinen der Formel

$$a \tilde{+} b = (a + b)(1 + \delta)\,, \qquad |\delta| \leqq \varepsilon\,. \tag{2}$$

Wenn die Maschine mit L Gleitkomma-Dualziffern rechnet, ist speziell $\varepsilon = (1 + c) \cdot 2^{-L}$; dabei ist c eine nichtnegative Maschinenkonstante (bei optimaler Rundung ist $c = 0$).

Auf Grund dieser Formel erhält WILKINSON für Verfahren I folgendes Ergebnis ($\tilde{s}$ ist der berechnete Wert für s):

$$\tilde{s} = \sum_{m=1}^{n} a_m (1 + \eta_m) = s + \sum_{m=1}^{n} a_m \eta_m \tag{I,2}$$

$$1 + \eta_m = \prod_{i=m}^{n} (1 + \varepsilon_i)\,, \qquad |\varepsilon_i| \leqq \varepsilon \quad \text{für} \quad i > 1\,, \quad \varepsilon_1 = 0\,, \tag{I,3}$$

wobei die Größen η_m durch

$$\begin{aligned} |\eta_{n-t+1}| &\leqq (1 + \varepsilon)^t - 1 \qquad (1 \leqq t \leqq n - 1) \\ |\eta_1| &\leqq (1 + \varepsilon)^{n-1} - 1 \end{aligned} \tag{I,4}$$

beschränkt sind; diese Schranken werden für $\varepsilon_i = \varepsilon$ (für $i > 1$) angenommen.

Für Verfahren II gilt: ist $\tilde{s}_{mk}$ der berechnete Wert für s_{mk}, so ist nach (2) und (II,1):

$$\tilde{s}_{mk} = (\tilde{s}_{2m-1,k-1} + \tilde{s}_{2m,k-1})(1 + \varepsilon_{mk}) \qquad \text{für} \quad 1 \leqq k \leqq t\,,$$
$$0 \leqq m \leqq 2^{t-k} - 1\,, \qquad |\varepsilon_{mk}| \leqq \varepsilon\,.$$

Durch vollständige Induktion erhält man daraus leicht das Ergebnis:

$$\tilde{s}_{mk} = \sum_{i=2^k(m-1)+1}^{2^k m} s_{i0} (1 + \delta_i^{(k)})\,, \qquad 1 + \delta_i^{(k)} = \prod_{L=1}^{k} \Big(1 + \varepsilon_{1+\left[\frac{i-1}{2^L}\right],L}\Big).$$

Für den berechneten Wert $\tilde{s} = \tilde{s}_{1t}$ der Summe s gilt also (unter Berücksichtigung von $s_{i0} = 0$ für $i > n$):

$$\tilde{s} = \sum_{m=1}^{n} a_m (1 + \eta_m) = s + \sum_{m=1}^{n} a_m \eta_m \tag{II,2}$$

$$1 + \eta_m = \prod_{i=1}^{t} \Big(1 + \varepsilon_{1+\left[\frac{m-1}{2^i}\right],i}\Big)\,, \qquad |\varepsilon_{ik}| \leqq \varepsilon\,, \tag{II,3}$$

woraus man sofort die Schranke

$$|\eta_i| \leqq (1 + \varepsilon)^t - 1 \qquad (1 \leqq i \leqq n) \tag{II,4}$$

erhält, die für $\varepsilon_{ik} = \varepsilon$ angenommen wird.

4. Rundungsfehler beim Kahan-Babuška-Verfahren (III)

Die berechneten Näherungswerte für s_m, w_m, s seien bzw. $\tilde{s}_m, \tilde{w}_m, \tilde{s}$. Aus (III,1) folgt jetzt mit (2):

$$\begin{aligned} \tilde{s}_0 &= \tilde{w}_0 = 0 \\ \tilde{s}_m &= (\tilde{s}_{m-1} + a_m)(1 + \varepsilon_m) \qquad m = 1, \ldots, n \\ \tilde{w}_m &= (1 + \alpha_m)\left(\tilde{w}_{m-1} + (1 + \beta_m)(a_m + (1 + \gamma_m)(\tilde{s}_{m-1} - \tilde{s}_m))\right) \qquad m = 1, \ldots, n \end{aligned} \tag{III,2}$$

mit $\alpha_m, \beta_m, \gamma_m, \varepsilon_m$, die betragsmäßig $\leqq \varepsilon$ sind. Da man annehmen kann, daß $z + 0 = z$ exakt berechnet wird, kann man voraussetzen:

$$\alpha_1 = \beta_1 = \gamma_1 = \varepsilon_1 = 0\,. \tag{III,2a}$$

Nun ist $q_m = \sum_{r=1}^{m} \left(\psi_r \prod_{i=r}^{m} \varphi_i\right)$ die Lösung der Differenzengleichung $q_m = \varphi_m (q_{m-1} + \psi_m)$ mit der Anfangsbedingung $q_0 = 0$, wie man durch Einsetzen feststellt.

Mit den Abkürzungen

$$\prod_{i=r}^{s} (1 + \varepsilon_i) = b_{rs}\,, \qquad (1 + \beta_r) \prod_{i=r}^{n} (1 + \alpha_i) = c_r \tag{III,3}$$

ist daher

$$\tilde{s}_m = \sum_{r=1}^{m} a_r b_{rm} \tag{III,4}$$

$$\tilde{w}_n = \sum_{r=1}^{n} c_1 \left(a_r + (1 + \gamma_r)(\tilde{s}_{r-1} - \tilde{s}_r)\right) \tag{III,5}$$

oder, unter Verwendung von (III,4):

$$\tilde{w}_n = \sum_{r=1}^{n} a_r \left\{c_r \left(1 - (1 + \gamma_r)(1 + \varepsilon_r)\right) + \sum_{s=r+1}^{n} c_s (1 + \gamma_s)(b_{r,s-1} - b_{rs})\right\}. \tag{III,6}$$

Also ist

$$\tilde{s}_n + \tilde{w}_n = \sum_{r=1}^{n} a_r (1 + \delta_r) \tag{III,7}$$

mit

$$\delta_r = 1 + b_{rn} - c_r \gamma_r (1 + \varepsilon_r) - c_r \varepsilon_r + \sum_{s=r+1}^{n} c_s (b_{r,s-1} - b_{rs})(1 + \gamma_s)\,. \tag{III,8}$$

Durch Umformen erhält man

$$\delta_r = -\varepsilon_r (c_r - 1) - c_r \gamma_r (1 + \varepsilon_r) + \sum_{s=r+1}^{n} (b_{r,s-1} - b_{rs})\left(c_s (\gamma_s + 1) - 1\right)$$

(wegen $b_{rn} = b_{rr} - \sum_{s=r+1}^{n} (b_{r,s-1} - b_{rs})$ und $b_{rr} = 1 + \varepsilon_r$) und weiter

$$-\delta_r = \varepsilon_r (c_r - 1) + c_r \gamma_r (1 + \varepsilon_r) + \sum_{s=r+1}^{n} \varepsilon_s b_{r,s-1} (c_s \gamma_s + c_s - 1)$$

$\left(\text{wegen } b_{rs} = (1 + \varepsilon_s) b_{r,s-1}\right)$, also

$$-\delta_r = \sum_{s=r}^{n} \varepsilon_s b_{r,s-1} (c_s \gamma_s + c_s - 1) + \gamma_r c_r\,. \tag{III,9}$$

Aus (III,3) folgt für $r > 1$:

$$\begin{aligned} |b_{rs}| &\leqq (1 + \varepsilon)^{s-r+1}\,, \qquad |c_s| \leqq (1 + \varepsilon)^{n-s+2}\,, \\ |c_s - 1| &\leqq (1 + \varepsilon)^{n-s+2} - 1 \end{aligned} \tag{III,10a}$$

und für $r = 1$ wegen (III,2a):

$$|b_{1s}| \leqq (1 + \varepsilon)^{s-1} \tag{III,10b}$$

Man erhält damit aus (III,9) die Abschätzung

$$\begin{aligned} |\delta_r| &\leqq (1 + \varepsilon)^{n-r+2} \varepsilon + \sum_{s=r}^{n} \varepsilon (1 + \varepsilon)^{s-r} \left(\varepsilon (1 + \varepsilon)^{n-s+2} + (1 + \varepsilon)^{n-s+2} - 1\right) \\ &= \varepsilon (1 + \varepsilon)^{n-r+2} + (n - r + 1)\, \varepsilon (1 + \varepsilon)^{n-r+3} - (1 + \varepsilon)^{n-r+1} + 1 \end{aligned}$$

für $r > 1$, also

$$|\delta_{n-t+1}| \leqq \big(\varepsilon(1+\varepsilon)^2 t + \varepsilon(1+\varepsilon) - 1\big)(1+\varepsilon)^t + 1 \qquad (1 \leqq t \leqq n-1) \tag{III,11a}$$

und

$$|\delta_1| \leqq \sum_{s=2}^{n} \varepsilon(1+\varepsilon)^{s-2}\big(\varepsilon(1+\varepsilon)^{n-s+2} + (1+\varepsilon)^{n-s+2} - 1\big)$$
$$= (n-1)\,\varepsilon(1+\varepsilon)^{n+1} - (1+\varepsilon)^{n-1} + 1\,,$$

also

$$|\delta_1| \leqq \big(\varepsilon(1+\varepsilon)^2(n-1) - 1\big)(1+\varepsilon)^{n-1} + 1\,. \tag{III,11b}$$

Die Schranken in (III,11) werden für $\alpha_i = \beta_i = \gamma_i = \varepsilon_i = \varepsilon$ $(i > 1)$ erreicht, können also i. a. nicht mehr verbessert werden.

Nun ist

$$\tilde{s} = (\tilde{s}_n + \tilde{w}_n)(1+\delta) = \left(s + \sum_{i=1}^{n} \alpha_i \delta_i\right)(1+\delta)\,,$$

also

$$\tilde{s} = s + \delta s + \sum_{i=1}^{n} a_i \eta_i \tag{III,12}$$

mit

$$\eta_i = -(\delta+1)\left(\gamma_r c_r + \sum_{s=r}^{n} \varepsilon_s b_{r,s-1}\big(c_s(\gamma_s+1) - 1\big)\right) \tag{III,13}$$

$$|\eta_{n-t+1}| \leqq \big(\varepsilon(1+\varepsilon)^2 t + \varepsilon(1+\varepsilon) - 1\big)(1+\varepsilon)^{t+1} + 1 + \varepsilon \qquad (1 \leqq t \leqq n-1)$$
$$|\eta_1| \leqq \big(\varepsilon(1+\varepsilon)^2(n-1) - 1\big)(1+\varepsilon)^n + 1 + \varepsilon \tag{III,14}$$
$$|\delta| \leqq \varepsilon\,.$$

5. Forderungen an eine Rechenmaschine. Ein Satz über Rundungsfehler

Zunächst werden einige Forderungen an die zu verwendende Rechenmaschine gestellt, die in der Praxis mehr oder weniger genau erfüllt sind. Zur Vorbereitung der Abschätzung für die Rundungsfehler beim Verfahren IV wird dann ein Satz über die Berechnung von $B - (A + B)$ für $|A| \leqq |B|$ in einer solchen Rechenmaschine bewiesen.

Über die Struktur der Maschinenzahlen wird folgende Annahme gemacht:

Voraussetzung 1: Maschinenzahlen sind genau die Zahl 0 und die Zahlen $x = \pm q \cdot Z^{-\alpha}$ mit $Z^{L-1} \leqq q < Z^L$, $q \in \mathbb{N}$, $\alpha \in \mathbb{G}$ *); dabei sind $Z > 1$ und L positive, ganzzahlige Maschinenkonstanten. In der Praxis können natürlich nur endlich viele α zugelassen werden. Alles Folgende bleibt jedoch richtig, wenn alle auftretenden Summen oder Differenzen eines dieser α als Exponent besitzen.

Hilfssatz 1: *Gilt Vor.* 1, *sind A und B Maschinenzahlen, gilt $A \geqq B \geqq 0$ und ist*

$$B = q\,Z^{-\alpha} \qquad (q \in \mathbb{N}\,, \quad Z^{L-1} \leqq q < Z^L\,, \quad \alpha \in \mathbb{G})\,,$$

so ist $A = r\,Z^{-\alpha}$ mit einem $r \in \mathbb{N}$, $r \geqq q$.

Beweis: Es ist $A \geqq B \geqq Z^{L-\alpha-1}$. Da A eine Maschinenzahl ist und da $A > 0$, ist $A = r_0\,Z^{-\beta}$ mit $r_0 \in \mathbb{N}$, $Z^{L-1} \leqq r_0 < Z^L$. Also ist $Z^{L-\beta} > r_0\,Z^{-\beta} = A \geqq Z^{L-\alpha-1}$ oder $L - \beta > L - \alpha - 1$. Das bedeutet aber, daß $-\beta = \gamma - \alpha$ mit $\gamma > 0$. Mit $r = r_0\,Z^{\gamma} (\in \mathbb{N})$ ist dann $A = r\,Z^{-\alpha}$. Aus $A \geqq B$ folgt schließlich $r \geqq q$, w. z. z. w.

Hilfssatz 2: *Gilt Vor.* 1, *ist A Maschinenzahl, $Z^{L-\alpha} > A \geqq Z^{L-1-\alpha}$, $\alpha \in \mathbb{G}$, so ist*

$$A = r\,Z^{-\alpha}\,, \qquad r \in \mathbb{N}\,, \qquad Z^{L-1} \leqq r < Z^L\,.$$

Beweis: Wegen $Z^{L-1-\alpha} = q\,Z^{-\alpha}$ mit $q = Z^{L-1}$ folgt aus Hilfssatz 1: $A = r\,Z^{-\alpha}$ mit $r \in \mathbb{N}$, $Z^{L-1} \leqq r$. Wegen $A < Z^{L-\alpha}$ gilt dann $r < Z^L$, w. z. z. w.

Für die Maschinenoperationen $\tilde{+}$ (Näherung für $+$) und $\tilde{-}$ (Näherung für $-$) seien folgende Voraussetzungen erfüllt:

Voraussetzung 2: Sind A und B Maschinenzahlen, gilt

(V 0) $\qquad A \geqq B \geqq 0$

und existiert ein $\alpha \in \mathbb{G}$ mit

(V 1) $\qquad A\,Z^{\alpha} \in \mathbb{G}\,, \qquad B\,Z^{\alpha} \in \mathbb{G}$

*) $\mathbb{G}$: Menge der ganzen Zahlen.

(V 2) $B < Z^{L-\alpha}$

(V 3) $A - B \leqq Z^{L-\alpha}$

so ist $A \,\tilde{-}\, B = A - B$.

Voraussetzung 3: Sind A und B Maschinenzahlen, gilt

$$A \geqq B \geqq 0\,, \qquad B \leqq Z^{L-\alpha}\,,$$

so ist

(U 1) $A \leqq A \,\tilde{+}\, B \leqq A + Z^{L-\alpha}$

(U 2) $A - Z^{L-\alpha} \leqq A \,\tilde{-}\, B \leqq A$

(U 3) $0 \leqq A \,\tilde{-}\, B$.

Bemerkung: Bei optimaler Rundung[2]) sind unter Annahme von Vor. 1 die Voraussetzungen 2 und 3 erfüllt.

Satz 1: *Es mögen Vor. 1—3 erfüllt sein. X, Y seien Maschinenzahlen mit $X \geqq Y > 0$. Dann gilt*

a) $(X \,\tilde{+}\, Y) \,\tilde{-}\, X = (X \,\tilde{+}\, Y) - X$

b) $X \,\tilde{-}\, (X \,\tilde{-}\, Y) = X - (X \,\tilde{-}\, Y)$.

Beweis: Nach Voraussetzung 1 ist

$$X = q\,Z^{-\gamma}\,, \qquad Z^{L-1} \leqq q < Z^{L}\,, \qquad (q \in \mathbb{N},\ \gamma \in \mathbb{G})\,, \tag{3}$$

also gilt

$$Y < Z^{L-\gamma}\,, \qquad Z^{L-\gamma-1} \leqq X < Z^{L-\gamma}\,. \tag{4}$$

a) Setze $A := X \,\tilde{+}\, Y$, $B := X$, $\alpha := \gamma$.

Nach (U 1) ist $A \geqq X$. Daher ist nach Hilfssatz 1:

$$A = r\,Z^{-\alpha}\,, \qquad r \geqq q\,, \qquad (r \in \mathbb{N})$$

$A \geqq X = B \geqq 0$ liefert (V 0), $A\,Z^{\alpha} = r \in \mathbb{G}$, $B\,Z^{\alpha} = q \in \mathbb{G}$ liefert (V 1), $B = X < Z^{L-\alpha}$ liefert (V 2).

Wegen $Y < Z^{L-\alpha}$ ist nach Vor. 3 (U 1):

$$A - B = (X \,\tilde{+}\, Y) - X \leqq (X + Z^{L-\alpha}) - X = Z^{L-\alpha}\,.$$

Also gilt auch (V 3) und damit ist nach Vor. 2: $(X \,\tilde{+}\, Y) \,\tilde{-}\, X = (X \,\tilde{+}\, Y) - X$.

b) Setze $A := X$, $B := X \,\tilde{-}\, Y$.

Nach (U 2), (U 3) gilt

$$0 \leqq B \leqq A = X < Z^{L-\gamma} \tag{5}$$

(V 0) ist daher erfüllt.

Es werden folgende Fälle unterschieden:

Fall α: $Z^{L-1-\gamma} \leqq Y < Z^{L-\gamma}$

Fall β: $Z^{L-1-\gamma} \leqq B < Z^{L-\gamma}$, $\quad Y < Z^{L-1-\gamma}$

Fall γ: $Z^{L-2-\gamma} \leqq B < Z^{L-1-\gamma}$, $\quad Y < Z^{L-1-\gamma}$

Fall δ: $Z^{L-2-\gamma} \leqq Y < Z^{L-1-\gamma}$, $\quad B < Z^{L-2-\gamma}$.

Die Möglichkeit $Y < Z^{L-2-\gamma}$, $B < Z^{L-2-\gamma}$ kann nicht eintreten, da daraus (im Widerspruch zu (4)) nach (U 2): $X - Z^{L-2-\gamma} \leqq B < Z^{L-2-\gamma}$, also $X < 2\,Z^{L-2-\gamma} \leqq Z^{L-1-\gamma}$ folgen würde.

Fall α: Setze $\alpha := \gamma$. Dann ist $Z^{L-1-\gamma} \leqq Y < Z^{L-\alpha}$, nach Hilfssatz 2 also

$$Y = r\,Z^{-\alpha}\,, \qquad Z^{L-1} \leqq r < Z^{L}\,, \qquad (r \in \mathbb{N})\,. \tag{6}$$

Nach Voraussetzung ist $X \geqq Y > 0$. Weiter ist $X\,Z^{\alpha} = q \in \mathbb{G}$, $Y\,Z^{\alpha} = r \in \mathbb{N}$, $Y < Z^{L-\alpha}$, $X - Y \leqq X < Z^{L-\alpha}$. Vor. 2 läßt sich daher auf X und Y anwenden und liefert

$$B = X \,\tilde{-}\, Y = X - Y = (q - r)\,Z^{-\alpha}\,.$$

Nun folgt (V 1) aus $A\,Z^{\alpha} = q \in \mathbb{G}$, $B\,Z^{\alpha} = q - r \in \mathbb{G}$, (V 2) aus $B \leqq A = X < Z^{L-\alpha}$, (V 3) aus

$$A - B = q\,Z^{-\alpha} - (q - r)\,Z^{-\alpha} = r\,Z^{-\alpha} \leqq Z^{L-\alpha} \quad \text{nach (6)}\,.$$

Fall β: Setze $\alpha := \gamma$. Dann ist $Z^{L-1-\alpha} \leqq B < Z^{L-\alpha}$, nach Hilfssatz 2 also

$$B = r\,Z^{-\alpha}\,, \qquad Z^{L-1} \leqq r < Z^{L}\,, \qquad (r \in \mathbb{N})\,.$$

---

[2]) $x \,\tilde{\pm}\, y$ = nächste bei $x \pm y$ gelegene Maschinenzahl; liegt $x \pm y$ genau in der Mitte zwischen den beiden nächsten Maschinenzahlen, so ist entweder stets $x \,\tilde{\pm}\, y$ = die betragsmäßig größere der beiden Maschinenzahlen oder stets $x \,\tilde{\pm}\, y$ = die betragsmäßig kleinere der beiden Maschinenzahlen.

(V 1) folgt aus $A\,Z^\alpha = q \in \mathfrak{G}$, $B\,Z^\alpha = r \in \mathfrak{G}$, (V 2) aus $B \leqq A = X < Z^{L-\alpha}$, (V 3) aus $A - B \leqq A = X < Z^{L-\alpha}$.

Fall γ: Setze $\alpha := \gamma + 1$. Dann ist $Z^{L-1-\alpha} \leqq B < Z^{L-\alpha}$, nach Hilfssatz 2 also

$$B = r\,Z^{-\alpha}, \qquad Z^{L-1} \leqq r < Z^L, \qquad (r \in \mathbb{N}). \tag{7}$$

(V 1) folgt aus $A\,Z^\alpha = q\,Z \in \mathfrak{G}$, $B\,Z^\alpha = r \in \mathfrak{G}$, (V 2) aus (7), (V 3) aus

$$A - B = X - (X \tilde{-} Y) \leqq X - (X - Z^{L-\alpha}) = Z^{L-\alpha}$$

wegen (U 2) und $Y \leqq Z^{L-\alpha}$.

Fall δ: Setze $\alpha := \gamma + 1$. Dann ist $Z^{L-1-\alpha} \leqq Y \leqq Z^{L-\alpha}$, nach Hilfssatz 2 also

$$Y = r\,Z^{-\alpha}, \qquad Z^{L-1} \leqq r \leqq Z^L, \qquad (r \in \mathbb{N}) \tag{8}$$

Nach (U 2) gilt: $X - Z^{L-\alpha} \leqq B < Z^{L-\alpha-1}$, also

$$X - Y \leqq Z^{L-\alpha} + Z^{L-\alpha-1} - Z^{L-\alpha-1} = Z^{L-\alpha}.$$

Wegen $X\,Z^\alpha = q\,Z \in \mathfrak{G}$, $Y\,Z^\alpha = r \in \mathfrak{G}$, $Y < Z^{L-\alpha}$ läßt sich Vor. 2 anwenden und man erhält:

$$B = X \tilde{-} Y = X - Y = (Z\,q - r)\,Z^{-\alpha}. \tag{9}$$

(V 1) folgt jetzt aus $A\,Z^\alpha = Z\,q \in \mathfrak{G}$, $B\,Z^\alpha = Z\,q - r \in \mathfrak{G}$, (V 2) aus (8), (9), (V 3) aus $A - B = X - (X - Y) = Y < Z^{L-\alpha}$. Also sind in allen vier Fällen die Bedingungen (V 1), (V 2), (V 3) erfüllt. Wegen (5) ist daher Vor. 2 anwendbar und liefert $X \tilde{-} (X \tilde{-} Y) = X - (X \tilde{-} Y)$, w. z. z. w.

Um aus diesem Satz Aussagen über $B \tilde{-} (A \tilde{+} B)$ zu gewinnen, werden folgende weitere Voraussetzungen getroffen (die bei optimaler Rundung wieder erfüllt sind):

Voraussetzung 4: Für alle Maschinenzahlen A gilt:

(a) $A \tilde{+} 0 = 0 \tilde{+} A = A$

(b) $A \tilde{-} A = 0$

(c) $0 \tilde{-} A = -A$.

Voraussetzung 5: Für alle positiven Maschinenzahlen X, Y gilt:

(d) $Y \tilde{+} X = X \tilde{+} Y$

(e) $Y \tilde{-} X = -(X \tilde{-} Y)$

(f) $(-X) \tilde{+} (-Y) = -(Y \tilde{+} X)$

(g) $(-X) \tilde{-} (-Y) = -(X \tilde{-} Y)$

(h) $X \tilde{+} (-Y) = X \tilde{-} Y$.

Satz 2: *Unter Vor.* 1–5 *gilt:*
Sind A, B *Maschinenzahlen mit* $|A| \leqq |B|$, *so ist*

$$B \tilde{-} (A \tilde{+} B) = B - (A \tilde{+} B).$$

Beweis: Ist $A = 0$, so ist $B \tilde{-} (A \tilde{+} B) = B \tilde{-} B = 0 = B - B = B - (A \tilde{+} B)$ nach Vor. 4.
Ist $B = 0$, so ist $B \tilde{-} (A \tilde{+} B) = B \tilde{-} A = 0 \tilde{-} A = -A = 0 - A = B - A = B - (A \tilde{+} B)$ nach Vor. 4.
Ist $A > 0$, $B > 0$, so ist mit $A = Y$, $B = X$:

$$B \tilde{-} (A \tilde{+} B) \overset{(e)}{=} -((Y \tilde{+} X) - X) \overset{(d)}{=} -((X \tilde{+} Y) \tilde{-} X) \overset{\text{Satz 1}}{=} -((X \tilde{+} Y) - X)$$
$$= X - (X \tilde{+} Y) \overset{(d)}{=} B - (A \tilde{+} B).$$

Ist $A > 0$, $B < 0$, so ist mit $A = Y$, $B = -X$:

$$B \tilde{-} (A \tilde{+} B) \overset{(h)}{=} (-X) \tilde{-} (Y \tilde{-} X) \overset{(c)}{=} (-X) \tilde{-} (-(X \tilde{-} Y)) \overset{(g)}{=} -(X \tilde{-} (X \tilde{-} Y))$$
$$\overset{\text{Satz 1}}{=} -(X - (X \tilde{-} Y)) \overset{(e)}{=} (-X) - (Y \tilde{-} X) \overset{(h)}{=} (-X) - (Y \tilde{+} (-X)) = B - (A \tilde{+} B).$$

Ist $A < 0$, $B > 0$, so ist mit $A = -Y$, $B = X$:

$$B \tilde{-} (A \tilde{+} B) \overset{(d)}{=} X \tilde{-} (X \tilde{+} (-Y)) \overset{(h)}{=} X \tilde{-} (X \tilde{-} Y) \overset{\text{Satz 1}}{=} X - (X \tilde{-} Y) \overset{(h)}{=} X - (X \tilde{+} (-Y)) \overset{(d)}{=} B - (A \tilde{+} B).$$

Ist schließlich $A < 0$, $B < 0$, so ist mit $A = -Y$, $B = -X$:

$$B \tilde{-} (A \tilde{+} B) \overset{(f)}{=} (-X) \tilde{-} (-(X \tilde{+} Y)) \overset{(g)}{=} -(X \tilde{-} (X \tilde{+} Y)) \overset{(e)}{=} (X \tilde{+} Y) \tilde{-} X \overset{\text{Satz 1}}{=} (X \tilde{+} Y) - X$$
$$= -X - (-(X \tilde{+} Y)) \overset{(f)}{=} B - (A \tilde{+} B).$$

Damit ist Satz 2 bewiesen.

6. Das verbesserte Kahan-Babuška-Verfahren (IV)

Um einfache Abschätzungen zu gewinnen, sei angenommen, daß die verwendete Maschine die Voraussetzungen 1 bis 5 erfüllt (also z. B. optimale Rundung besitzt). Für andere Maschinen kann man die erhaltene Abschätzung als näherungsweise gültig ansehen. Die Bezeichnung der in der Maschine berechneten Werte sei wie in Nummer 4. Dann ist für $|a_m| \leqq |\tilde{s}_{m-1}|$ unter Verwendung von (2):

$$a_m \tilde{+} (\tilde{s}_{m-1} \tilde{-} \tilde{s}_m) = a_m \tilde{+} (\tilde{s}_{m-1} \tilde{-} (a_m \tilde{+} \tilde{s}_{m-1})) \overset{\text{Satz 2}}{=} a_m \tilde{+} (\tilde{s}_{m-1} - (a_m \tilde{+} \tilde{s}_{m-1}))$$
$$= a_m \tilde{+} (\tilde{s}_{m-1} - \tilde{s}_m) = (1 + \beta_m)(a_m + \tilde{s}_{m-1} - \tilde{s}_m)$$

und für $|a_m| > |\tilde{s}_{m-1}|$:

$$\tilde{s}_{m-1} \tilde{+} (a_m \tilde{-} \tilde{s}_m) = \tilde{s}_{m-1} \tilde{+} (a_m \tilde{-} (\tilde{s}_{m-1} \tilde{+} a_m)) \overset{\text{Satz 2}}{=} \tilde{s}_{m-1} \tilde{+} (a_m - (\tilde{s}_{m-1} \tilde{+} a_m))$$
$$= \tilde{s}_{m-1} \tilde{+} (a_m - \tilde{s}_m) = (1 + \beta_m)(\tilde{s}_{m-1} + a_m - \tilde{s}_m)\,.$$

Daher ist in beiden Fällen:

$$\tilde{s}_m = (1 + \varepsilon_m)(\tilde{s}_{m-1} + a_m) \tag{IV,2}$$
$$\tilde{w}_m = (1 + \alpha_m)\big(\tilde{w}_{m-1} + (1 + \beta_m)(a_m + \tilde{s}_{m-1} - \tilde{s}_m)\big)\,. \tag{IV,3}$$

Das sind aber gerade die Formeln (III,2) für $\gamma_m = 0$ $(m \geqq 1)$. (Man kann also eine Verbesserung gegenüber Verfahren III erwarten. Dies ist der Grund für die Einführung dieses Verfahrens.) Daher folgen aus (IV,2, 3) die folgenden Formeln ((III,7, 9, 10) für $\gamma_m = 0$):

$$s' := \tilde{s}_n + \tilde{w}_n = \sum_{r=1}^{n} a_r (1 + \delta_r) \tag{IV,4}$$
$$\delta_r = -\sum_{s=r}^{n} \varepsilon_s b_{r,s-1}(c_s - 1) \tag{IV,5}$$
$$|b_{rs}| \leqq (1+\varepsilon)^{s-r+1}\,, \qquad |c_s - 1| \leqq (1+\varepsilon)^{n-s+2} - 1\,, \qquad |b_{1s}| \leqq (1+\varepsilon)^{s-1}\,. \tag{IV,6}$$

Daraus erhält man für $r > 1$:

$$|\delta_r| \leqq \sum_{s=r}^{n} \varepsilon (1+\varepsilon)^{s-r}\big((1+\varepsilon)^{n+2-s} - 1\big) = \varepsilon (1+\varepsilon)^{n-r+2}(n-r+1) - (1+\varepsilon)^{n-r+1} + 1\,,$$

also

$$|\delta_{n-t+1}| \leqq (\varepsilon(1+\varepsilon)\,t - 1)(1+\varepsilon)^t + 1 \qquad (1 \leqq t \leqq n-1) \tag{IV,7a}$$

und für $r = 1$:

$$|\delta_1| \leqq \sum_{s=2}^{n} \varepsilon (1+\varepsilon)^{s-2}\big((1+\varepsilon)^{n-s+2} - 1\big) = \varepsilon (n-1)(1+\varepsilon)^n - (1+\varepsilon)^{n-1} + 1\,,$$

also

$$|\delta_1| \leqq \big(\varepsilon(1+\varepsilon)(n-1) - 1\big)(1+\varepsilon)^{n-1} + 1\,. \tag{IV,7b}$$

Nun ist $\tilde{s} = (\tilde{s}_n + \tilde{w}_n)(1+\delta) = s'(1+\delta) = \left(s + \sum_{i=1}^{n} a_i \delta_i\right)(1+\delta)$, also ist

$$\tilde{s} = s + \delta s + \sum_{i=1}^{n} a_i \eta_i \tag{IV,8}$$
$$\eta_r = -(1+\delta)\sum_{s=r}^{n} \varepsilon_s b_{r,s-1}(c_s - 1) \tag{IV,9}$$
$$|\eta_{n-t+1}| \leqq \big(\varepsilon(1+\varepsilon)\,t - 1\big)(1+\varepsilon)^{t+1} + 1 + \varepsilon \qquad (1 \leqq t \leqq n-1)$$
$$|\eta_1| \leqq \big(\varepsilon(1+\varepsilon)(n-1) - 1\big)(1+\varepsilon)^n + 1 + \varepsilon \tag{IV,10}$$
$$|\delta| \leqq \varepsilon\,.$$

Wie man aus (IV,9) sieht, werden auch die Schranken in (IV,10) angenommen, und zwar für $\alpha_m = \beta_m = \varepsilon_m = \varepsilon$ $(m > 1)$.

7. Vereinfachung der Abschätzungen für kleine ε

Es ist

$$\varepsilon - \frac{\varepsilon^2}{2} \leqq r \leqq \varepsilon \qquad \text{für} \quad r = \ln(1+\varepsilon)\,. \tag{10}$$

Daher ist für Verfahren I nach (I,4):

$$|\eta_{n-t+2}| \leqq (1+\varepsilon)^{t-1} - 1 = \varepsilon\,(t-1) + r^2 t^2 \frac{e^{rt} - r\,t - 1}{(r\,t)^2} - t(\varepsilon - r) - \varepsilon\left((1+\varepsilon)^{t-1} - 1\right)$$

$$\leqq \varepsilon\,(t-1) + \varepsilon^2 t^2 \frac{e^k - k - 1}{k^2} \qquad \text{für} \qquad \varepsilon\, t \leqq k\,,$$

ebenso

$$|\eta_1| \leqq (1+\varepsilon)^{n-1} - 1 \leqq \varepsilon\,(n-1) + \varepsilon^2 n^2 \frac{e^k - k - 1}{k^2} \qquad \text{für} \qquad \varepsilon\, n \leqq k\,,$$

also

$$\begin{aligned} &|\eta_i| \leqq \varepsilon\,(t-1) + \varepsilon^2 t^2 \frac{e^k - k - 1}{k^2} \qquad (\varepsilon\, n \leqq k) \\ &t = n - i + 2 \quad \text{für} \quad i > 1\,, \qquad t = n \quad \text{für} \quad i = 1\,. \end{aligned} \tag{I,5}$$

Für Verfahren II gilt (mit $u = t - 1$):

$$|\eta_i| \leqq (1+\varepsilon)^{u+1} - 1 = \varepsilon\,(u+1) + r^2 u^2 \frac{e^{ru} - r\,u - 1}{(r\,u)^2} + \varepsilon\, r\, u \frac{e^{ru} - 1}{r\,u} - u\,(\varepsilon - r)\,,$$

$$|\eta_i| \leqq \varepsilon\, t + \varepsilon^2 \left((t-1)^2 \frac{e^k - k - 1}{k^2} + (t-1)\frac{e^k - 1}{k}\right) \qquad \left(\varepsilon\,(t-1) \leqq k\right) \tag{II,5}$$

$$t \leqq \log_2 n + 1\,.$$

Für Verfahren III ist nach (III,14):

$$|\eta_{n-t+2}| \leqq \left(\varepsilon\,(1+\varepsilon)^2 t - \varepsilon^2 (1+\varepsilon) - 1\right)(1+\varepsilon)^t + 1 + \varepsilon$$

$$= \varepsilon + r^2 t^2 \frac{e^{rt}(r\,t - 1) + 1}{(r\,t)^2} + \varepsilon\, r\, t \frac{5}{2} e^{rt} + \varepsilon^2 e^{rt}(\ldots)$$

$$\leqq \varepsilon + \varepsilon^2 \left(t^2 \frac{e^k (k-1) + 1}{k^2} + t \frac{5}{2} e^k\right) \qquad \left(\varepsilon\, t \leqq k \leqq \frac{4}{9}\,,\ \varepsilon \leqq \frac{1}{2}\right),$$

denn $\dfrac{e^x (x-1) + 1}{x^2} = \sum\limits_{i=0}^{\infty} \dfrac{i+1}{(i+2)!} x^i$ ist für $x > 0$ monoton wachsend, und aus (10) erhält man durch eine einfache Abschätzung $(\ldots) \leqq 0$ für $\varepsilon\, t \leqq \frac{4}{9}$, $\varepsilon \leqq \frac{1}{2}$. Aus (III,14) sieht man, daß $|\eta_1|$ durch die Schranke von $|\eta_2|$ beschränkt wird. Daher gilt:

$$\begin{aligned} &|\eta_i| \leqq \varepsilon + \varepsilon^2 \left(t^2 \frac{e^k (k-1) + 1}{k^2} + t \frac{5}{2} e^k\right) \left(\varepsilon\, t \leqq k \leqq \frac{4}{9}\,,\ \varepsilon \leqq \frac{1}{2}\right) \\ &t = n - i + 2 \quad \text{für} \quad i > 1\,, \qquad t = n \quad \text{für} \quad i = 1\,. \end{aligned} \tag{III,15}$$

Für Verfahren IV erhält man aus (IV,10):

$$|\eta_{n-t+2}| \leqq \left(\varepsilon\,(1+\varepsilon)(t-1) - 1\right)(1+\varepsilon)^t + 1 + \varepsilon$$

$$= r^2 t^2 \frac{e^{rt}(r\,t - 1) + 1}{(r\,t)^2} + \varepsilon\, r\, t \frac{e^{rt}(3\,r\,t - 2) + 2}{2\,r\,t} + \varepsilon^2 e^{rt}(\ldots)$$

$$\leqq \varepsilon^2 \left(t^2 \frac{e^k (k-1) + 1}{k^2} + t \frac{e^k (3\,k - 2) + 2}{2\,k}\right) \left(\varepsilon\, t \leqq k \leqq 1\,,\ \varepsilon \leqq \frac{1}{2}\right),$$

da $\dfrac{e^x (3\,x - 2) + 2}{x^2} = \sum\limits_{i=0}^{\infty} \dfrac{3\,i + 1}{(i+1)!} x^i$ für $x > 0$ monoton wächst und da $(\ldots) \leqq 0$ für $\varepsilon\, t \leqq 1$, $\varepsilon \leqq \frac{1}{2}$.

Wieder wird $|\eta_1|$ durch die Schranke von $|\eta_2|$ beschränkt, also:

$$\begin{aligned} &|\eta_i| \leqq \varepsilon^2 \left(t^2 \frac{e^k (k-1) + 1}{k^2} + t \frac{e^k (3\,k - 2) + 2}{2\,k}\right) \left(\varepsilon\, t \leqq k \leqq 1\,,\ \varepsilon \leqq \frac{1}{2}\right), \\ &t = n - i + 2 \quad \text{für} \quad i > 1\,, \qquad t = n \quad \text{für} \quad i = 1\,. \end{aligned} \tag{IV,11}$$

Nun gilt für alle vier Verfahren eine Gleichung der Form

$$\tilde{s} = s + \delta s + \sum_{i=1}^{n} a_i\, \eta_i \tag{11}$$

($|\delta| \leqq \varepsilon$ für Verfahren III, IV; $\delta = 0$ für Verfahren I, II). Daraus lassen sich zwei einfache Formeln für den absoluten Fehler $s - \tilde{s}$ ableiten. Denn aus (11) folgt:

$$|s - \tilde{s}| \leqq |\delta|\,|s| + \sum_{i=1}^{n} |a_i|\,|\eta_i|;$$

also

$$|s - \tilde{s}| \leqq |\delta|\,|s| + \left(\operatorname*{Max}_{1\leqq i\leqq n} |\eta_i|\right)\left(\sum_{i=1}^{n} |a_i|\right) \tag{12}$$

$$|s - \tilde{s}| \leqq |\delta|\,|s| + \left(\sum_{i=1}^{n} |\eta_i|\right)\left(\operatorname*{Max}_{1\leqq i\leqq n} |a_i|\right). \tag{13}$$

Formel (12) ist nützlich, wenn alle a_i dasselbe Vorzeichen haben, Formel (13), wenn alle a_i von derselben Größenordnung sind. Auf diesen Formeln baut die Diskussion der Verfahren in diesem und dem nächsten Abschnitt auf.

Im folgenden wird $\varepsilon n \leqq k = 1/3$ vorausgesetzt, was in der Praxis wohl meist erfüllt ist. Dann gilt:

$$\frac{e^k - 1}{k} < \frac{6}{5}, \quad \frac{e^k(k-1)+1}{k^2} < \frac{3}{4}, \quad \frac{e^k - k - 1}{k^2} < \frac{3}{5}, \quad \frac{5}{2} e^k < \frac{7}{2}, \quad \frac{e^k(3k-2)+2}{2k} < 1. \tag{14}$$

Für die einzelnen Verfahren erhält man aus (12), (13), damit folgende Formeln:

Verfahren I: Aus (I,5) erhält man mit (14):

$$\operatorname*{Max}_{1\leqq i\leqq n} |\eta_i| \leqq \varepsilon(n-1) + \frac{3}{5}\varepsilon^2 n^2$$

$$\sum_{i=1}^{n} |\eta_i| \leqq \varepsilon(n-1) + \frac{3}{5}\varepsilon^2 n^2 + \sum_{t=2}^{n}\left(\varepsilon(t-1) + \frac{3}{5}\varepsilon^2 t^2\right),$$

$$\sum_{i=1}^{n} |\eta_i| \leqq \varepsilon\,\frac{n^2+n-2}{2} + \varepsilon^2\,\frac{2n^3+9n^2+n-6}{10},$$

$$\leqq \varepsilon\,\frac{n^2+n-2}{2} + \varepsilon^2\,\frac{n^3+5n^2}{5}$$

also

$$|s - \tilde{s}| \leqq \left(\varepsilon(n-1) + \frac{2}{5}\varepsilon^2 n^2\right)\sum_{i=1}^{n} |a_i| \quad \left(\varepsilon n \leqq \frac{1}{3}\right) \tag{I,6}$$

$$|s - \tilde{s}| \leqq \left(\varepsilon\,\frac{n^2+n-2}{2} + \varepsilon^2\,\frac{n^3+5n^2}{5}\right)\operatorname*{Max}_{1\leqq i\leqq n} |a_i| \quad \left(\varepsilon n \leqq \frac{1}{3}\right) \tag{I,7}$$

Verfahren II: Aus (II,2), (II,5), (14) folgt unmittelbar:

$$|s - \tilde{s}| \leqq \left(\varepsilon(1+\log_2 n) + \frac{3}{5}\varepsilon^2 \log_2 n\,(\log_2 n + 2)\right)\sum_{i=1}^{n} |a_i| \quad \left(\varepsilon n \leqq \frac{1}{3}\right), \tag{II,6}$$

$$|s - \tilde{s}| \leqq \left(\varepsilon(1+\log_2 n) + \frac{3}{5}\varepsilon^2 \log_2 n\,(\log_2 n + 2)\right) n \operatorname*{Max}_{1\leqq i\leqq n} |a_i| \quad \left(\varepsilon n \leqq \frac{1}{3}\right). \tag{II,7}$$

Verfahren III: Nach (III,15) ist

$$\operatorname*{Max}_{1\leqq i\leqq n} |\eta_i| \leqq \varepsilon + \varepsilon^2\left(\frac{3}{4}n^2 + \frac{7}{2}n\right)$$

$$\sum_{i=1}^{n} |\eta_i| \leqq \varepsilon + \varepsilon^2\left(\frac{3}{4}n^2 + \frac{7}{2}n\right) + \sum_{t=2}^{n}\left(\varepsilon + \varepsilon^2\left(\frac{3}{4}t^2 + \frac{7}{2}t\right)\right)$$

$$= \varepsilon n + \varepsilon^2\left(\frac{1}{4}n^3 + \frac{23}{8}n^2 + \frac{43}{8}n - \frac{17}{4}\right) \leqq \varepsilon n + \varepsilon^2\left(\frac{1}{4}n^3 + 3n^2 + 4n\right),$$

also

$$|s - \tilde{s}| \leqq \varepsilon|s| + \left(\varepsilon + \varepsilon^2\left(\frac{3}{4}n^2 + \frac{7}{2}n\right)\right)\sum_{i=1}^{n} |a_i| \quad \left(\varepsilon n \leqq \frac{1}{3}\right) \tag{III,16}$$

$$|s - \tilde{s}| \leqq \varepsilon|s| + \left(\varepsilon n + \varepsilon^2\left(\frac{1}{4}n^3 + 3n^2 + 4n\right)\right)\operatorname*{Max}_{1\leqq i\leqq n} |a_i| \quad \left(\varepsilon n \leqq \frac{1}{3}\right). \tag{III,17}$$

Verfahren IV: Aus (IV,11) folgt

$$\operatorname*{Max}_{1\leq i\leq n} |\eta_i| \leqq \varepsilon^2 \left(\frac{3}{4} n^2 + n\right)$$

$$\sum_{i=1}^{n} |\eta_i| \leqq \varepsilon^2 \left(\frac{3}{4} n^2 + n\right) + \sum_{i=2}^{n} \varepsilon^2 \left(\frac{3}{4} t^2 + t\right) = \varepsilon^2 \left(\frac{1}{4} n^3 + \frac{17}{8} n^2 + \frac{25}{8} n - \frac{11}{4}\right)$$

$$\leqq \varepsilon^2 \left(\frac{1}{4} n^3 + \frac{5}{2} n^2 + n\right),$$

also

$$|s - \tilde{s}| \leqq \varepsilon\, |s| + \varepsilon^2 \left(\frac{3}{4} n^2 + n\right) \sum_{i=1}^{n} |a_i| \qquad \left(\varepsilon\, n \leqq \frac{1}{3}\right) \qquad \text{(IV,12)}$$

$$|s - \tilde{s}| \leqq \varepsilon\, |s| + \varepsilon^2 \left(\frac{1}{4} n^3 + \frac{5}{2} n^2 + n\right) \operatorname*{Max}_{1\leq i\leq n} |a_i| \qquad \left(\varepsilon\, n \leqq \frac{1}{3}\right). \qquad \text{(IV,13)}$$

Die Ergebnisse ((I,6, 7), (II,6, 7), (III,16, 17), (IV,12, 13)) sind in Tabelle 1 zusammengefaßt.

Bemerkung: Der ε-Anteil der Verfahrensfehler von I, II, III wurde — in etwas anderer Form — schon von Babuška [3] angegeben.

Tabelle 1. Fehlerkoeffizienten der Summierungsverfahren

$$\left\{\begin{aligned} &|s - \tilde{s}| \leqq A\, \varepsilon |s| + (B\, \varepsilon + C\, \varepsilon^2) \sum_{i=1}^{n} |a_i| \\ &|s - \tilde{s}| \leqq A\, \varepsilon |s| + (D\, \varepsilon + E\, \varepsilon^2) \operatorname*{Max}_{1\leq i\leq n} |a_i| \end{aligned}\right\} \text{für } \varepsilon\, n \leqq \frac{1}{3} \qquad \begin{matrix}(15)\\(16)\end{matrix}$$

Verf. Nr.	A	$B(n)$	$C(n)$	$D(n)$	$E(n)$
I	0	$n - 1$	$0{,}6\, n^2$	$0{,}5\, n^2 + 0{,}5\, n - 1$	$0{,}2\, n^3 + n^2$
II	0	$1 + \log_2 n$	$0{,}6\, (\log_2 n)^2 + 1{,}2 \log_2 n$	$n\, (1 + \log_2 n)$	$0{,}6\, n \log_2 n$
III	1	1	$0{,}75\, n^2 + 3{,}5\, n$	n	$0{,}25\, n^3 + 3\, n^2 + 4\, n$
IV	1	0	$0{,}75\, n^2 + n$	0	$0{,}25\, n^3 + 2{,}5\, n^2 + n$

8. Diskussion der Ergebnisse

8.1. Zunächst sollen alle a_i dasselbe Vorzeichen haben. Dann ist mit der Abkürzung

$$s_{\text{abs}} := \sum_{i=1}^{n} |a_i| \qquad (17)$$

$s_{\text{abs}} = |s|$ und der relative Fehler ist nach (15)

$$\left|\frac{s - \tilde{s}}{s}\right| \leqq (A + B)\, \varepsilon + C\, \varepsilon^2 . \qquad (18)$$

Solange $n < \varepsilon^{-1/2}$ ist, spielt der Term $C\, \varepsilon^2$ in (18) gegenüber $(A + B)\, \varepsilon$ keine Rolle. Das erste Verfahren hat dann einen maximalen Fehler, der proportional zur Zahl n der Summanden wächst. Im zweiten Verfahren hat man trotz gleichen Rechenaufwands einen maximalen Fehler, der nur mit dem Logarithmus von n wächst (ist z. B. $n = 10^6$, so ist der Verfahren-I-Fehler $10^6\, \varepsilon$, der Verfahren-II-Fehler nur $20\, \varepsilon$!!). Wenn daher die Zahl der Summanden von vornherein bekannt ist, empfiehlt es sich, anstelle von Verfahren I stets Verfahren II zu benutzen. (Bei der Summierung einer unendlichen Reihe ist das natürlich nicht möglich.)

Beim Kahan-Babuška-Verfahren (III) ist der maximale Fehler fast konstant. Mit etwa $2\, \varepsilon$ liegt er in derselben Größenordnung wie der Fehler bei einer Normalverfahrenrechnung mit doppelter Genauigkeit (ε^2 statt ε), bei der das Ergebnis auf einfache Genauigkeit gerundet wird. Da aber der Rechenaufwand für doppelte Genauigkeit nur doppelt so groß ist wie der für einfache, lohnt sich Verfahren II nur dann, wenn auf der Maschine keine Möglichkeit, mit doppelter Genauigkeit zu rechnen, besteht. Dann ist das Verfahren ein guter Ersatz dafür (das gilt nur für $s_{\text{abs}} \approx s$; vgl. unten!). Die Fehlerschranke von Verfahren IV hat ebenfalls die Größenordnung ε. Die geringfügige Verbesserung gegenüber III lohnt den höheren Aufwand zur Berechnung der Summe jedoch nicht.

Ist $\varepsilon^{-1} \leqq n^2 \ll \varepsilon^{-1} \log_2 (\varepsilon^{-1})$, so wachsen die Schranken in (18) zwar quadratisch in n bzw. $\log_2 n$, aber das Obengesagte bleibt qualitativ richtig. Für $n^2 \approx \varepsilon^{-1} \log_2 (\varepsilon^{-1})$ sind Verfahren II und III etwa gleich genau; für $n^2 \geqq 2\, \varepsilon^{-1} \log_2 (\varepsilon^{-1})$ ist schließlich Verfahren II das genaueste Verfahren (genauer als III und IV), da

$$0{,}6\, (\log_2 n)^2 + 1{,}2 \log_2 n \leqq n \qquad \text{für alle } n ,$$

$$\log_2 n \leqq \frac{3}{4} n^2\, \varepsilon \qquad \text{für} \qquad n^2 \geqq 2\, \varepsilon^{-1} \log_2 (\varepsilon^{-1})$$

$\left(\frac{\log_2 t}{t}\right.$ ist für $t \geqq 3$ monoton fallend und $\leqq \frac{3}{2}\varepsilon$ für $t = 2\,\varepsilon^{-1}\log_2(\varepsilon^{-1})$, also ist $\log_2 n = \frac{1}{2} n^2 \frac{\log_2 n^2}{n^2} \leqq \frac{3}{4}$ für $\left. n \geqq 2\,\varepsilon^{-1}\log_2(\varepsilon^{-1})\right)$.

8.2. Es sei $s_{\text{abs}} \approx |s|$. Dann gilt (18) angenähert. Daher lassen sich alle obigen Aussagen näherungsweise übertragen. Insbesondere ist Verfahren III fast so gut wie Verfahren IV, für kleine n ist Verfahren III (IV) das beste, für große n ist Verfahren II am besten geeignet. Verfahren I ist wieder das schlechteste Verfahren von allen.

8.3. Es sei $s_{\text{abs}} \gg |s|$. Jetzt kann man das Glied $A\,\varepsilon\,|s|$ vernachlässigen und man erhält

$$|s - \tilde{s}| \leqq (B + C\,\varepsilon)\,(s_{\text{abs}}\,\varepsilon)\,. \tag{19}$$

8.3.1. Ist s_{abs} so groß, daß $s_{\text{abs}} \cdot \varepsilon$ mindestens die Größenordnung von $|s|$ hat (etwa wenn s fast Null ist), so ist der Näherungswert $\tilde{s}$ i. a. nur dann brauchbar, wenn $B = 0$ ist, also wenn s nach IV berechnet wird. In den anderen Fällen kann eine so starke Auslöschung auftreten, daß nur noch eine oder keine Ziffer von $\tilde{s}$ mehr richtig ist.

8.3.2. Auch wenn der Extremfall 8.3.1 nicht vorliegt, bestimmt die Größe von B zunehmend die Fehlerschranke. Für Verfahren I ist der maximale relative Fehler jetzt $\gg n\,\varepsilon$, für große n ist das Verfahren unbrauchbar. Verfahren II hat jetzt einen Fehler von ungefähr $\varepsilon_{\text{abs}}\log_2 n\; s_{\text{abs}}/|s|$, der in nicht zu ungünstigen Fällen etwa die Größenordnung $n\,\varepsilon$ haben kann. Dieses Verfahren ist also noch so gut wie Verfahren I im günstigsten Fall, also bedingt verwendbar. Für das KAHAN-BABUŠKA-Verfahren ist die Schranke ungefähr $\varepsilon\, s_{\text{abs}}/|s|$ (für kleine und mittlere n), wird also in der Mehrzahl der Fälle (z. B. Summation absolut konvergenter Reihen mit alternierenden Vorzeichen) noch gute Ergebisse liefern. Das verbesserte KAHAN-BABUŠKA-Verfahren hat jedoch eine Schranke von größenordnungsmäßig $n^2\,\varepsilon^2\,\frac{s_{\text{abs}}}{|s|}$ für den relativen, $n^2\,\varepsilon^2\,s_{\text{abs}}$ für den absoluten Fehler, liefert also für nicht allzu große n (z. B. für $n \ll \varepsilon^{-1/2}$) einen Maximalfehler, der in der Größenordnung von ε liegt! Hier wird also die Auslöschung fast vollständig kompensiert. In diesem Fall wird sich daher der gegenüber I fünffache Rechenaufwand praktisch immer lohnen (wenn man nicht mit doppelter Genauigkeit rechnen kann).

8.4. Wenn s_{abs} nicht sehr groß gegenüber $|s|$ ist, so ist wieder Verfahren II dem ersten Verfahren vorzuziehen. Ob man Verfahren III (oder IV) verwendet, wird von der benötigten Genauigkeit der Summe abhängen. Verfahren III und IV dürften dann etwa gleichwertig sein, der Vorteil größerer Genauigkeit bei IV wird durch den größeren Rechenaufwand aufgewogen.

Das hier Gesagte schließt natürlich nicht aus, daß in Einzelfällen Verfahren I bessere Ergebnisse liefert als Verfahren II oder III; oder das Verfahren I auch einmal für ein kleines n besser ist als Verfahren III; Beispiele dafür werden im nächsten Abschnitt gegeben.

9. Beispiele

Für die ersten drei Beispiele sei eine L-ziffrige Gleitkomma-Dualmaschine mit optimaler Rundung[3]) zugrunde gelegt; eine genaue 1 in der $(L+1)^{\text{ten}}$. Dualstelle soll abgerundet werden. Nach der Bemerkung in Abschnitt 3 ist daher $\varepsilon = 2^{-L}$.

a) Es sei $n = 7$,

$$a_1 = (2^{-1} + 2^{1-L} + 2^{-L})\,2^k\,,$$
$$a_2 = a_3 = a_4 = a_5 + (2^{-1})\,2^{k-1-L}$$
$$a_6 = (2^{-1})\,2^{k+2}$$
$$a_7 = -\,(2^{-1} + 2^{-3} + 2^{-L})\,2^{k+2}\,.$$

Es ist $s = 0$. Tabelle 2 gibt die Zwischenwerte für die Summation nach Verfahren I, III und IV wieder, Tabelle 3 die für Verfahren II.

Tabelle 2. Zwischenwerte zu Beispiel a)

m	$\tilde{s}_m$	$(\tilde{w}_m\text{III})$	$\tilde{w}_m(\text{IV})$	Ergebnisse
1	$(2^{-1} + 2^{1-L} + 2^{-L})\,2^k$	0	0	$\tilde{s}_{\text{I}} = 0$
2	$(2^{-1} + 2^{1-L} + 2^{-L})\,2^k$	$(2^{-1})\,2^{k-1-L}$	$(2^{-1})\,2^{k-1-L}$	$\tilde{s}_{\text{III}} = (2^{-1})\,2^{k+1-L}$
3	$(2^{-1} + 2^{1-L} + 2^{-L})\,2^k$	$(2^{-1})\,2^{k-L}$	$(2^{-1})\,2^{k-L}$	$\tilde{s}_{\text{IV}} = 0$
4	$(2^{-1} + 2^{1-L} + 2^{-L})\,2^k$	$(2^{-1} + 2^{-2})\,2^{k-L}$	$(2^{-1} + 2^{-2})\,2^{k-L}$	
5	$(2^{-1} + 2^{1-L} + 2^{-L})\,2^k$	$(2^{-1})\,2^{k+1-L}$	$(2^{-1})\,2^{k+1-L}$	
6	$(2^{-1} + 2^{-3} + 2^{-L})\,2^{k+2}$	$(2^{-1})\,2^{k+1-L}$	0	
7	0	$(2^{-1})\,2^{k+1-L}$	0	

[3]) siehe Fußnote [2]).

Tabelle 3. Zwischenwerte zu Beispiel a)

m	$\tilde{s}_{m0}$	$\tilde{s}_{m1}$	$\tilde{s}_{m2}$	$\tilde{s}_{m3}$
1	$(2^{-1} + 2^{1-L})\,2^k$	$(2^{-1} + 2^{1-L} + 2^{-L})\,2^k$	$(2^{-1} + 2^{1-L} + 2^{-L})\,2^k$	$-(2^{-1})\,2^{k+1}$
2	$(2^{-1})\,2^{k-1-L}$	$(2^{-1})\,2^{k-L}$	$-(2^{-1} + 2^{2-L})\,2^k$	
3	$(2^{-1})\,2^{k-1-L}$	$(2^{-1})\,2^{k+2}$		
4	$(2^{-1})\,2^{k-1-L}$	$-(2^{-1} + 2^{-3} + 2^{-L})\,2^{k+2}$		
5	$(2^{-1})\,2^{k-1-L}$			
6	$(2^{-1})\,2^{k+2}$			
7	$-(2^{-1} + 2^{-3} + 2^{-L})\,2^{k+2}$			
8	0			

Ergebnis: $\tilde{s}_{II} = -(2^{-1})\,2^{k+1-L}$

Für die absoluten Fehler erhält man

$$|s - \tilde{s}_I| = |s - \tilde{s}_{IV}| = 0\,, \qquad |s - \tilde{s}_{II}| = |s - \tilde{s}_{III}| = 2^{k-L} = \frac{1}{2}\,\varepsilon \max_{1 \leq i \leq 7} |a_i|\,.$$

Die in Tabelle 1 angegebenen Schranken überschätzen den wirklichen Fehler bei Verfahren III um das Doppelte, bei Verfahren II um das Achtfache. Der nach I berechnete Wert ist sogar genau. Trotz der wesentlich schlechteren Schranke in Tab. 1 ist also Verfahren I hier genauer als Verfahren II oder III. Verfahren IV liefert das exakte Ergebnis.

b) Es sei $n = 4$, $a_1 = 1$, $a_2 = 2^{-2} + 2^{-1-L}$, $a_3 = -1$, $a_4 = -(2^{-2} + 2^{-1-L})$. Wie man leicht nachprüft, ist jetzt

$$s = 0\,, \qquad \tilde{s}_{II} = \tilde{s}_{IV} = 0\,, \qquad \tilde{s}_I = \tilde{s}_{III} = -2^{-L-1}\,,$$

also Verfahren II genauer als Verfahren I oder III. Verfahren IV gibt wieder das exakte Ergebnis.

c) Sei $n = 3$, $a_1 = 1$, $a_2 = 2^{L+1} + 2^L$, $a_3 = -(2^{L+1} + 2^L)$. Jetzt ist

$$s = 1\,, \qquad \tilde{s}_1 = \tilde{s}_{II} = 0\,, \qquad \tilde{s}_{III} = \tilde{s}_{IV} = 1\,,$$

also Verfahren I und II ungenauer als Verfahren III oder IV.

Meistens werden von allen Verfahren ungenaue Ergebnisse geliefert, z. B. stets, wenn s keine Maschinenzahl ist. Ist aber s als Maschinenzahl darstellbar, so ist oft $\tilde{s}_{IV} = s$; vgl. dazu Beispiel d), Tab. 4, 5. ($n = 7$; die a_m und die Zwischenwerte sind in den Tabellen angegeben.)

Tabelle 4. Zwischenwerte zu Beispiel d)

m	a_m	$\tilde{s}_m$	$\tilde{w}_m$(III)	$\tilde{w}_m$(IV)	Ergebnisse
1	0.5555_{10^0}	0.5555_{10^0}	0	0	$\tilde{s}_I = 0.7780_{10^1}$
2	0.5555_{10^1}	0.6111_{10^1}	$-0.1000_{10^{-2}}$	$-0.5000_{10^{-3}}$	$\tilde{s}_{III} = 0.7774_{10^1}$
3	0.5555_{10^0}	0.6667_{10^1}	$-0.1500_{10^{-2}}$	$-0.1000_{10^{-2}}$	$\tilde{s}_{IV} = 0.7777_{10^1}$
4	0.5555_{10^1}	0.7223_{10^1}	$-0.2000_{10^{-2}}$	$-0.1500_{10^{-2}}$	
5	0.5555_{10^2}	0.6277_{10^2}	$-0.2000_{10^{-2}}$	$+0.1500_{10^{-2}}$	
6	0.5555_{10^0}	0.6333_{10^2}	$-0.6500_{10^{-2}}$	$-0.3000_{10^{-2}}$	
7	-0.5555_{10^2}	0.7780_{10^1}	$-0.6500_{10^{-2}}$	$-0.3000_{10^{-2}}$	

Tabelle 5. Zwischenwerte zu Beispiel d)

m	$\tilde{s}_{m0}$	$\tilde{s}_{m1}$	$\tilde{s}_{m2}$	$\tilde{s}_{m3}$
1	0.5555_{10^0}	0.6111_{10^1}	0.7222_{10^1}	0.7782_{10^1}
2	0.5555_{10^1}	0.1111_{10^1}	0.5600_{10^0}	
3	0.5555_{10^0}	0.5611_{10^2}		
4	0.5555_{10^1}	-0.5555_{10^2}		
5	0.5555_{10^2}			
6	0.5555_{10^0}			
7	-0.5555_{10^2}			
8	0			

Ergebnis: $\tilde{s}_{II} = 0.7782_{10^1}$

Die Werte wurden unter Annahme einer Dezimalmaschine mit vier geltenden Ziffern und der üblichen Rundung berechnet. Das genaue Ergebnis ist $s = 0.7777_{10^1} = \tilde{s}_{IV}$. *Wie zu erwarten war,* nimmt jetzt die Genauigkeit der Verfahren in der Reihenfolge I, II, III, IV zu.

An den Tabellen 2 und 4 sieht man gut, woran es liegt, daß Verfahren III i. a. schlechter ist als Verfahren IV. Immer wenn ein großes a_m zu einem kleinen s_{m-1} addiert wird, berechnet der Korrektor w_m des KAHAN-BABUŠKA-Verfahrens nicht mehr, wie geplant, den Rundungsfehler, sondern einen Wert, der mit dem Rundungsfehler nicht mehr viel zu tun hat. Die im verbesserten KAHAN-BABUŠKA-Verfahren eingebaute Abfrage vertauscht hier sozusagen die Rollen von a_m und s_{m-1} und berechnet dadurch — nach Satz 2 des fünften Abschnitts — den *richtigen* Fehler.

NICKEL [3] berechnete auf einer Maschine mit $\varepsilon = 2^{-30}$ die Summe $\sum_{i=1}^{2^t} \frac{1}{i}$ für $0 \leqq t \leqq 14$ nach Verfahren I und III (= IV in diesem Fall). Für $s = \sum_{i=1}^{2^{12}} \frac{1}{i}$ z. B. erhält er

$$|s - \tilde{s}_{\mathrm{I}}| \approx 25 \cdot 10^{-8}, \qquad |s - \tilde{s}_{\mathrm{IV}}| \leqq 0.5 \cdot 10^{-8}.$$

Unter Verwendung von $s < 9$ ergibt sich aus Tabelle 1:

$$|s - \tilde{s}_{\mathrm{I}}| \leqq 3500 \cdot 10^{-8}, \qquad |s - \tilde{s}_{\mathrm{IV}}| \leqq 1.7 \cdot 10^{-8}.$$

Der Fehler von Verfahren I wird also durch die Schranke in Tab. 1 um das 140fache, bei Verfahren IV mindestens um das dreifache überschätzt. Jedoch liegt die Tab. 1-Schranke für Verfahren IV immer noch niedriger als der maschinell berechnete Fehler von Verfahren I.

Das letzte Beispiel soll wieder auf einer L-ziffrigen Dualmaschine gerechnet werden. Diesmal soll die Rundung jedoch durch Abschneiden der $(L + i)$-ten Ziffern $(i \geqq 1)$ geschehen $(\varepsilon = 2^{1-L})$. Nun sei $n = 2^t$, $2 \leqq t \leqq L - 1$,

$$a_1 = 1, \qquad a_2 = 2^{-L}, \ldots, a_{2^i+1} = \cdots = a_{2^{i+1}} = 2^{-L} + 2^{i-2L} \qquad (1 \leqq i \leqq t - 1).$$

Man findet leicht:

$$\tilde{s} = 1 + 2^{t-L} - 2^{-L} + \frac{1}{3} 2^{-2L} (2^{2t} - 4),$$

$$\tilde{s}_{\mathrm{I}} = 1,$$

$$\tilde{s}_{\mathrm{II}} = 1 + 2^{t-L} - 2^{1-L} \qquad \left(t > \frac{L}{2} + 1\right)$$

$$\tilde{s}_{\mathrm{II}} = 1 + 2^{t-L} - 2^{1-L} + \frac{1}{3} \cdot 2^{-2L}(2^{2t} - 2^{L+r}) \quad \left(t > \frac{L}{2} + 1; r = 2 \text{ für gerade } L, r = 1 \text{ für ungerade } L\right),$$

$$\tilde{s}_{\mathrm{III}} = 1 + 2^{t-L} - 2^{1-L}.$$

Man sieht unmittelbar, daß Verfahren II für $n^2 \geqq 8\,\varepsilon^{-1}$ genauer ist als Verfahren III (hier = Verfahren IV); das ist in guter Übereinstimmung mit der in Abschnitt 8 abgeleiteten Bedingung $n^2 \geqq \varepsilon^{-1} \log_2 \varepsilon^{-1}$. Daß $\tilde{s}_{\mathrm{II}}$ für kleine n hier $= \tilde{s}_{\mathrm{III}}$ ist, folgt daraus, daß wegen der speziellen Form der Summanden bei Verfahren II nur die $\tilde{s}_{1i}$ gerundet sind.

Für die Fehler erhält man

$$|\tilde{s}_{\mathrm{I}} - s| = 2^{t-L} + \frac{1}{3} (2^{t-L})^2 - 2^{-L} - \frac{1}{3} 2^{2-2L} = \frac{1}{2} (n - 1)\,\varepsilon + \left(\frac{1}{12} n^2 - \frac{1}{3}\right) \varepsilon^2,$$

$$|\tilde{s}_{\mathrm{II}} - s| = 2^{-L} + \frac{1}{3} 2^{-2L} (2^{2t} - 4) = \frac{1}{2} \varepsilon + \left(\frac{1}{12} n^2 - \frac{1}{3}\right) \varepsilon^2 \qquad (\text{für } n^2 < 8\,\varepsilon^{-1}),$$

$$|\tilde{s}_{\mathrm{II}} - s| = \left(\frac{1}{3} 2^r + 1\right) 2^{-L} - \frac{1}{3} 2^{2-2L} \leqq \frac{7}{6} \varepsilon - \frac{1}{3} \varepsilon^2 \qquad (\text{für } n^2 \geqq 8\,\varepsilon^{-1}),$$

$$|\tilde{s}_{\mathrm{III}} - s| = 2^{-L} + \frac{1}{3} 2^{-2L} (2^{2t} - 4) = \frac{1}{2} \varepsilon + \left(\frac{1}{12} n^2 - \frac{1}{3}\right) \varepsilon^2.$$

Die Fehlerschranken aus Tab. 1 überschätzen hier den Fehler bei Verfahren I um den Faktor 2, bei Verfahren II um den Faktor $2 \log_2 \varepsilon^{-1}$, bei Verfahren III um den Faktor 4 (da $s \approx 1$). Der hohe Fehler (im Vergleich zur Schranke) bei Verfahren I und III und der kleine Fehler bei Verfahren II rühren von der speziellen Form der a_i her; im allgemeinen werden die Fehler bei Verfahren I und III kleiner, bei Verfahren II größer sein. Trotzdem gibt dieses Beispiel die Unterschiede der benutzten Summierungsverfahren gut wieder, wie sie in Abschnitt 8.1 diskutiert wurden.

Literatur

1 BABUŠKA, I., Numerical Stability in Mathematical Analysis, Information Processing 68, Amsterdam, 11–22 (1969).

2 KAHAN, W., Further Remarks on Reducing Truncation Errors, Comm. ACM, 8, 40 (1965).

3 NICKEL, K., Das Kahan-Babuškasche Summierungsverfahren in Triplex-ALGOL 60, ZAMM 50, 369–373 (1970).

4 WILKINSON, J. H., Rounding Errors in Algebraic Processes, London 1963; dt. Übersetzung: Rundungsfehler, Berlin-Heidelberg-New York 1969.

Eingereicht am 6. 3. 1973

Anschrift: ARNOLD NEUMAIER, 75 Karlsruhe 41, Kieselweg 1, BRD

KLEINE MITTEILUNGEN

ZAMM 54, 53 – 55 (1974)

W. Wilhelmi

Ein Algorithmus zur Lösung eines inversen Eigenwertproblems*)

1. Aufgabenstellung

Viele technische Entwurfsaufgaben lassen sich auf folgendes Problem zurückführen:

Es sei

$$A = A_c + \sum_{k=1}^{q} B_k K C_k \tag{1}$$

eine quadratische Matrix n-ter Ordnung, wobei A_c eine konstante Matrix und K eine (s,s)-Diagonalmatrix mit variablen Diagonalelementen $k_1, \ldots, k_s$ sind.

B_k und C_k sind konstante (n, s)- bzw. (s, n)-Matrizen. Die Parameter $k_1, \ldots, k_s$ sollen so bestimmt werden, daß die Matrix A vorgegebene Zahlen $\lambda_1, \ldots, \lambda_r$ $(r \leqq n)$ als Eigenwerte hat. Diese Zahlen seien voneinander verschieden. Komplexe Zahlen λ_j sollen in konjugierten Paaren vorgegeben werden.

Diese Aufgabenstellung ist ein inverses Eigenwertproblem, das sowohl den additiven $(s = n,\ q = 1,\ B_1 = C_1 = I)$ als auch den multiplikativen $(s = n,\ q = 1,\ A_c = 0,\ B_1 = I)$ Fall umfaßt. In [2], [3], [4] werden Lösungsmethoden für additive und multiplikative Probleme angegeben. [2] und [4] setzen A_c als symmetrisch voraus, während in [3] die Vorgabe komplexer λ_j nicht behandelt wird. Mit den r vorgegebenen λ_j und dem s-dimensionalen Vektor

$$k = \begin{bmatrix} k_1 \\ \vdots \\ k_s \end{bmatrix}$$

der zu bestimmenden Parameterwerte k_1 sowie r noch zu bestimmenden Vektoren y_j mit der unitären Norm 1 wird folgende Funktion gebildet:

$$Q(k, y_j) = \sum_{j=1}^{r} y_j^* (A - \lambda_j I)^* (A - \lambda_j I)\, y_j$$

$$= \sum_{j=1}^{r} y_j^* H_j y_j\,; \quad \|y_j\| = 1 \tag{2}$$

In (2) sind die Matrix A entsprechend (1) bzw. die im allgemeinen positiv semidefiniten Matrizen

$$H_j(k) = (A - \lambda_j I)^* (A - \lambda_j I)\,; \quad j = 1, \ldots, r \tag{3}$$

von den Parametern k_1 abhängig.

Es ist $Q(k, y_j) = 0$ dann und nur dann, wenn gilt

$$A(k)\, y_j = \lambda_j y_j\,; \quad j = 1, \ldots, r \tag{4}$$

d. h. wenn $k_1, \ldots, k_s$ eine Lösung der inversen Eigenwertaufgabe ist. Für $s < r$ wird eine Lösung im allgemeinen nicht existieren. Für $s > r$ sind im allgemeinen keine isolierten Lösungen vorhanden. In Abschnitt 4 werden Bedingungen für die Existenz von Lösungen angegeben.

Die Bestimmung einer existierenden Lösung $\tilde{k}$ von $Q(k, y_j) = 0$ läßt sich als Minimierungsaufgabe darstellen:

$$Q(\tilde{k}, \tilde{y}_j) = \min Q(k, y_j)$$

$$k \in R_s\,; \quad y_j \in U_n, \quad \|y_j\| = 1\,; \quad j = 1, \ldots, r\,. \tag{5}$$

Dies ergibt sich daraus, daß jeder Summand von Q und damit auch Q gemäß Definition (2) stets nichtnegativ ist.

2. Lösungsalgorithmus

2.1. *Iterationsvorschrift*

Es wird abwechselnd auf dem $(r \times n)$-dimensionalen Raum der Vektoren y_j und dem s-dimensionalen Raum der Parameter k_l minimiert. Wegen des unabhängigen Eingehens der y_j in die Summanden von $Q(k, y_j)$ zerfällt der erstgenannte Schritt in r n-dimensionale Minimierungsaufgaben. Diese lassen sich einzeln wie folgt formulieren: Finde den zum kleinsten Eigenwert von H_j gehörigen Eigenvektor! Damit lautet der Iterationsalgorithmus

1. Wähle Anfangswerte für die Parameter $k_1^{(0)}, \ldots, k_s^{(0)}$. Setze $\nu = 0$.
2. Berechne $H_j^{(\nu)}$ aus $k^{(\nu)}$ nach (1) und (3) für $j = 1, \ldots, r$.
3. Bestimme den zum kleinsten Eigenwert $\alpha_j^{(\nu)}$ von $H_j^{(\nu)}$ gehörigen Eigenvektor $y_j^{(\nu)}$ für $j = 1, \ldots, r$.
4. Bestimme $k^{(\nu+1)}$ so, daß $Q(k^{(\nu+1)}, y_j^{(\nu)}) = \min Q(k, y_j^{(\nu)})$.
5. Erhöhe ν um 1 und gehe nach 2.

2.2. *Praktische Durchführung*

2.2.1. Schritt 3

Zur Bestimmung des zum kleinsten Eigenwertes gehörigen Eigenvektors wird das bekannte Verfahren der gebrochenen Iteration [5], [6] angewendet. Die gebrochene Iteration braucht bei komplexer Matrix H_j (als Folge von $Re\,(\lambda_j) \neq 0$) nur für eine der Zahlen λ_j aus dem vorgegebenen konjugierten Paar durchgeführt zu werden, da die entsprechenden iterierten Vektoren ebenfalls konjugiert komplex sind.

Zu Beginn dürfen beliebige Vektoren gewählt werden. Als Anfangswerte für die nächste Durchführung des Schrittes 3 mittels gebrochener Iteration dienen die im vorhergehenden Durchlauf bestimmten Vektoren $y_1^{(\nu-1)}, \ldots, y_r^{(\nu-1)}$, wodurch der Rechenaufwand besonders bei höheren Iterationsstufen gering gehalten wird. Wegen der Konvergenz der gebrochenen Iteration bricht sie bei passender Abbrechbedingung nach endlich vielen Wiederholungen ab. Dabei ist zu beachten, daß der gesuchte Eigenvektor nur approximiert wird.

Die Abweichung

$$y_j^{(\nu)*} H_j^{(\nu)} y_j^{(\nu)} - \alpha_j^{(\nu)} \geqq 0\,; \quad j = 1, \ldots, r \tag{6}$$

ist bei gegebenen $k_1^{(\nu)}, \ldots, k_s^{(\nu)}$ um so kleiner, je größer die Zahl der gebrochenen Iterationen ist. Um Komplikationen zu vermeiden, die bei singulärer oder fast singulärer Matrix $H_j^{(\nu)}$ eintreten können, wird empfohlen, die Iteration auf $H_j^{(\nu)} + dI$ $(d > 0)$ anzuwenden.

2.2.2. Schritt 4

Der zu minimierende Ausdruck läßt sich durch Einsetzen von (1) in (2) wie folgt errechnen:

$$Q(k, y_j^{(\nu)}) = Q(0, y_j^{(\nu)}) +$$

$$+ 2\, k^T \sum_{j=1}^{r} \left[\sum_{k=1}^{q} D(C_k \bar{y}_j^{(\nu)})\, B_k^T \right] (A_c - \lambda_j I)\, y_j^{(\nu)} \times$$

$$\times\, k^T \sum_{j=1}^{r} \left[\sum_{k=1}^{q} D(C_k \bar{y}_j^{(\nu)})\, B_k^T \right] \left[\sum_{k=1}^{q} B_k D(C_k y_j^{(\nu)}) \right] k$$

$$= Q(0, y_j^{(\nu)}) + 2\, k^T v^{(\nu)} + k^T M^{(\nu)} k\,. \tag{7}$$

In (7) ist $D(x)$ eine (s, s)-Diagonalmatrix, die auf dem i-ten Diagonalplatz die i-te Komponente des s-dimensionalen Vektors x enthält.

Die notwendige Bedingung für das Minimum von $Q(k, y_j^{(\nu)})$ lautet

$$\frac{1}{2} \frac{\partial}{\partial k} Q(k, y_j^{(\nu)}) = M^{(\nu)} k + v^{(\nu)} = 0\,. \tag{8}$$

Diese Bedingung ist auch hinreichend, weil die Funktion Q nach unten beschränkt ist und damit ein Minimum besitzt, für das (8) notwendig erfüllt ist. Wenn $M^{(\nu)}$ singulär sein sollte, dann bildet die allgemeine Lösung von (8) einen einfach zusammenhängenden Bereich, in dem $Q(k, y_j^{(\nu)})$ einen

---

*) Begründung und Anwendung des Verfahrens sind in der Dissertation [1] des Verfassers enthalten.

konstanten Wert hat, nämlich den Minimalwert, da in diesem Bereich das Minimum liegen muß. Damit ist jede singuläre Lösung von (8) brauchbar. Unter der getroffenen Annahme, daß komplexe λ_j und die entsprechenden Vektoren $\boldsymbol{y}_j^{(\nu)}$ in konjugierten Paaren auftreten, wird das lineare Gleichungssystem (8) für die Parameter $k_1^{(\nu+1)}, \ldots, k_s^{(\nu+1)}$ reell.

3. Konvergenzbetrachtung

Aus der Iterationsvorschrift folgt

$$\text{für Schritt 3: } Q(\boldsymbol{k}^{(\nu)}, \boldsymbol{y}_j^{(\nu)}) \leqq Q(\boldsymbol{k}^{(\nu)}, \boldsymbol{y}_j^{(\nu-1)}) \tag{9}$$

$$\text{für Schritt 4: } Q(\boldsymbol{k}^{(\nu+1)}, \boldsymbol{y}_j^{(\nu)}) \leqq Q(\boldsymbol{k}^{(\nu)}, \boldsymbol{y}_j^{(\nu)})\,. \tag{10}$$

Hierbei ist (9) bei beliebigem Abbruch der gebrochenen Iteration gültig, da die Folge der iterierten RAYLEIGH-Quotienten einer positiv definiten Matrix monoton fällt.

Die Folge $Q(\boldsymbol{k}^{(\nu)}, \boldsymbol{y}_j^{(\nu)})$ fällt wegen (9) und (10) monoton. Da sie nach unten durch Null beschränkt ist, konvergiert sie gegen einen Grenzwert. Es existiert mindestens ein Häufungspunkt $(\boldsymbol{k}, \tilde{\boldsymbol{y}}_j)$ der Iterationsfolge. Aus der Art der Iterationsdurchführung ergibt sich

$$Q(\tilde{\boldsymbol{k}}, \tilde{\boldsymbol{y}}_j) = \min_{\boldsymbol{y}_j} Q(\boldsymbol{y}_j, \tilde{\boldsymbol{k}}) \tag{11}$$

$$Q(\tilde{\boldsymbol{k}}, \tilde{\boldsymbol{y}}_j) = \min_{\boldsymbol{k}} Q(\tilde{\boldsymbol{y}}_j, \boldsymbol{k})\,. \tag{12}$$

Für die Gültigkeit von (11) und (12) muß natürlich die genaue Bestimmung der Eigenvektoren in Schritt 3 vorausgesetzt werden.

4. Existenz von Lösungen

Für die Existenz und Eindeutigkeit der Lösungen inverser Eigenwertprobleme sind keine einfachen hinreichenden und notwendigen Bedingungen bekannt, die sich auf die betrachtete Aufgabenstellung anwenden lassen [7].

Aus allen Spaltenvektoren der Matrizen B_k und allen Zeilenvektoren der Matrizen C_k $(k = 1, 2, \ldots, q)$ wählen wir alle linear unabhängigen aus, die dann die Basis der Darstellung aller Vektoren der entsprechenden Matrizen bilden. Es sind dies höchstens je n an der Zahl. Der h-te Spalten- oder Zeilenvektor von B_k bzw. C_k läßt sich dann mittels der Vektoren $\boldsymbol{v}_{kh}$ bzw. $\boldsymbol{w}_{kh}$ seiner kontravarianten Koordinaten wie folgt darstellen:

$$\boldsymbol{b}_{kh} = B\,\boldsymbol{v}_{kh}; \quad \boldsymbol{c}_{kh}^T = C^T \boldsymbol{w}_{kh}\,. \tag{13}$$

Die Matrizen B und C sind aus den gewählten Basisvektoren spalten- bzw. zeilenweise zusammengesetzt. Bezeichnet man die aus den Vektoren $\boldsymbol{v}_{kh}$ und $\boldsymbol{w}_{kh}$ als h-te Spalten gebildeten konstanten Matrizen mit V_k bzw. W_k, so lautet der Ausdruck für A folgendermaßen:

$$A = A_c + B\left(\sum_{k=1}^{q} V_k\,K\,W_k^T\right)C = A + B\,N\,C\,. \tag{14}$$

Die Elemente der Matrix N sind homogene lineare Formen in den Parametern k_l. Der Rang der Koeffizientenmatrix des entsprechenden linearen Gleichungssystems für die k_l sei $\tilde{r} \leqq s$.

Für die Existenz der Lösung des inversen Eigenwertproblems bei r beliebig vorgegebenen Eigenwerten ist notwendig

$$r \leqq \tilde{r}\,. \tag{15}$$

Die Erfüllung der Bedingung (15) bedeutet nämlich, daß in N genau $\tilde{r}$ Elemente beliebig bestimmt werden können. Dann sind die Eigenwerte der Matrix A von höchstens $\tilde{r}$ Parametern abhängig und folglich auch nur höchstens $\tilde{r}$ von ihnen beliebig vorgebbar.

Ohne Beweis wird folgende Bedingung angegeben [1]. Für die Existenz der Lösung des inversen Eigenwertproblems bei $\tilde{r}$ beliebig vorgegebenen Eigenwerten ist hinreichend, daß jede der Matrizen

$$\begin{array}{l}[B,\ A_c B,\ A_c{}^2 B, \ldots, A_c^{n-1}\ B] \\ [C^T,\ C^T A_c^T,\ C^T A_c^{T\,2}, \ldots, C^T A_c^{T\,n-1}]\end{array} \tag{16}$$

den Rang n hat.

5. Beispiel

Der entwickelte Algorithmus wurde in ALGOL-60 notiert. Zur Illustration wird hier ein Anwendungsbeispiel aus dem Gebiet der automatischen Steuerung [1] dargestellt.

$$n = 5, \quad q = 1, \quad r = 3;$$

$$A_c = \begin{bmatrix} -3{,}50 & 0 & 0 & 8{,}04 & 0 \\ 1{,}00 & -5{,}0 & 0 & 0 & 0 \\ 0 & -3{,}05 & -0{,}71 & 0 & 0 \\ 0 & 0 & 1{,}00 & -9{,}79 & 0 \\ 0 & 0 & 0 & -1{,}00 & 0 \end{bmatrix}$$

$$B_1 = \begin{bmatrix} 8{,}04 & 0 & 0 \\ 0 & 0 & 0 \\ 0 & 3{,}05 & 3{,}05 \\ 0 & 0 & 0 \\ 0 & 0 & 0 \end{bmatrix}; \quad C_1^T = \begin{bmatrix} 0 & 0 & 0 \\ -1{,}0 & 0 & 0 \\ 0 & 0 & 0 \\ 0 & -1{,}0 & 0 \\ 0 & 0 & -1{,}0 \end{bmatrix}$$

Es wurden folgende Eigenwerte vorgegeben $(s = 3)$:

$$\lambda_1 = -\,1{,}0 - i\sqrt{3{,}0}$$

$$\lambda_2 = -\,1{,}0 + i\sqrt{3{,}0}$$

$$\lambda_3 = -\,3{,}0\,.$$

Das Bild stellt den Verlauf der iterierten Parameterwerte dar, wobei ihre Anfangswerte zu Null angenommen wurden.

6. Diskussion

Aus (11) und (12) folgt nicht notwendig, daß die Iterationsfolge zur Lösung des inversen Eigenwertproblems nach (5) führt, selbst wenn eine Lösung existiert. Die berechneten Werte erfüllen nur die notwendigen Bedingungen (11) und (12). Unter Berücksichtigung der Näherungsfehler bei Schritt 3 liegt der betrachtete Häufungspunkt der Folge nicht einmal an dem durch (11) und (12) bezeichneten stationären Punkt von Q, sondern nur in einer gewissen Umgebung, die um so kleiner ist, je genauer die Eigenvektoren bestimmt werden.

Erweist sich, daß die Eigenwertgleichungen nicht erfüllt werden, was durch einen von Null verschiedenen Grenzwert von Q angezeigt wird, so müssen andere Anfangswerte $k_1^{(0)}, \ldots, k_s^{(0)}$ gewählt werden. Dieser Fall trat bei bisher behandelten Beispielen, für die Lösungen existierten, nicht auf.

Zur Beschleunigung der Konvergenz bei großen ν eignet sich eine Extrapolationsmethode, die auf der Vermutung basiert, daß nach einer gewissen Anzahl von Iterationsschritten ein monotoner Verlauf eintritt (s. Bild 1).

Dazu wird zur Berechnung von $A^{(\nu+1)}$ bzw. $H_j^{(\nu+1)}$ nicht $\boldsymbol{k}^{(\nu+1)}$, sondern

$$\boldsymbol{k}^{(\nu)} + \vartheta(\boldsymbol{k}^{(\nu+1)} - \boldsymbol{k}^{(\nu)}) \tag{17}$$

benutzt. Dadurch wird sich die Rechenzeit auf ungefähr den ϑ-ten Teil verringern, wobei ϑ nicht so groß gewählt werden darf, daß die Iteration instabil wird.

Bei Anwendung des Algorithmus auf konkrete Entwurfsaufgaben ist zu beachten, daß nur r Eigenwerte eingestellt werden, die übrigen $n - r$ sich jedoch unkontrolliert verschieben.

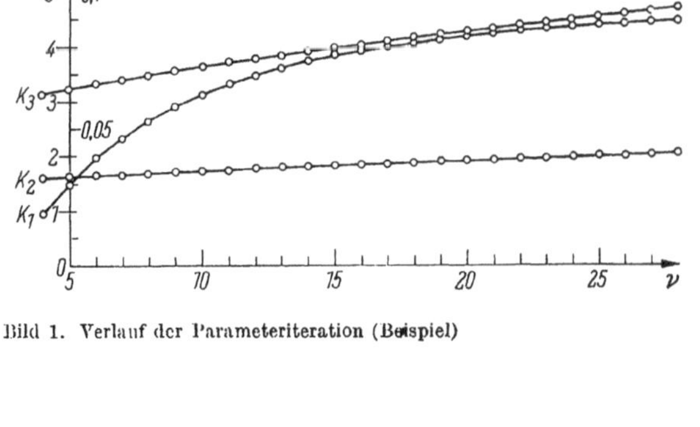

Bild 1. Verlauf der Parameteriteration (Beispiel)

Der Algorithmus zur Bestimmung der Parameter ist auch zur Lösung des teilweisen Eigenwertproblems für die Matrix A_c geeignet. Ein einzelner Eigenwert (sowie der dazugehörige Eigenvektor) wird bestimmt, indem Real- und Imaginärteil als Parameter angesehen werden, die Eigenwertgleichung durch Umformung auf die Form (4) gebracht und $\lambda_1 = 0$ vorgegeben wird.

Literatur

1 Wilhelmi, W., Beitrag zum Entwurf linearer zeitinvarianter Systeme mit unvollständiger Rückführung des Zustandes, Dissertation an der TU Dresden, Fakultät für Datenverarbeitung 1971.
2 Downing, A. C., Householder, A. S., Some inverse characteristic value problems, Journal of the ACM **3**, S. 203–207 (1956).
3 Uhlig, J., Ein Iterationsverfahren für ein inverses Eigenwertproblem endlicher Matrizen, ZAMM **40**, S. 123–125 (1960).
4 Hadeler, K. P., Newton-Verfahren für inverse Eigenwertaufgaben, Numerische Mathematik **12**, S. 35–39 (1968).
5 Zurmühl, R., Matrizen und ihre technische Anwendung, 4. Auflage, Berlin-Göttingen-Heidelberg 1964, Springer-Verlag.
6 Faddejev, D. K., Faddejewa, V. N., Numerische Methoden der linearen Algebra, 1. Auflage, Berlin 1964, VEB Deutscher Verlag der Wissenschaften.
7 Hadeler, K. P., Existenz- und Eindeutigkeitssätze für inverse Eigenwertaufgaben mit Hilfe des topologischen Abbildungsgrades, Archive for Rational Mechanics and Analysis **42**–4, S. 317–322 (1971).

Eingereicht am 13. 1. 1972, rev. Fassung am 16. 4. 1973

Anschrift: Dr. W. Wilhelmi, 108 Berlin, Leipziger Str. 47, DDR

ZAMM 54, 55–57 (1974)

D. Bhatia

Duality for Quadratic Programming in Complex Space

1. Introduction

Recently Rani [4] has considered symmetric dual quadratic programs and established duality theorems. The object of this note is to extend the duality theorems for the same class of programming problems in complex space over arbitrary polyhedral cones. This seems to be an extension over the results stated by Abrams and Ben Israel [5] in the sense that in their formulation, matrix D appears both in the constraint set as well as in the objective function. Here we wish to point out that this type of inter-dependence through the matrix D is not at all necessary.

In the last section some particular cases are discussed and in particular it is shown that results obtained by Sharma and Kaul [3] fall out as a particular case of these results.

2. Dual problems

Consider the following dual pair:

Problem I (Primal):

Minimize

$$f(u, v) = \operatorname{Re}\left(\frac{1}{2} u^* Du + \frac{1}{2} v^* Cv + p^* v\right),$$

subject to

$$A_1 v + A_2 u + b \in S, \tag{2.1}$$

$$v \in T. \tag{2.2}$$

Problem II (Dual):

Maximize

$$g(\xi, \eta, \omega) = \operatorname{Re}\left(-\frac{1}{2}\omega^* D\omega - \frac{1}{2}\xi^* C\,\xi - b^* \eta\right),$$

subject to

$$-A_1^* \eta + C\,\xi + p \in T^*, \tag{2.3}$$

$$\eta \in s^*, \tag{2.4}$$

$$A_2^* \eta - Dw = 0, \tag{2.5}$$

where C and D are positive semidefinite hermitian matrices of orders $n \times n$ and $m \times m$ respectively, A_1 and A_2 are $m \times n$ and $m \times m$ complex matrices respectively. u, b, w and η are m-dimensional complex vectors. Also v, p and ζ are n-dimensional complex vectors. S, T denote polyhedral cones with vertex 0 and S^*, T^* their dual cones. (* over a letter is used to denote the conjugate transpose). For proving duality theorems we shall be requiring the following lemmas:

Lemma 1: [3] If A is a positive semidefinite hermitian matrix of order n, z_1 and z_2 are any $n \times 1$ vectors, then

$$\operatorname{Re}(z_1^* A z_1 + z_2^* A z_2) \geqq 2 \operatorname{Re}(z_1^* A z_2).$$

Lemma 2: If (u_0, v_0) is an optimal solution for problem I, then it is optimal for the linear program I':

Problem I': Minimize

$$F(u, v) = \operatorname{Re}(u_0^* Du + v_0^* C v + p^* v),$$

subject to $A_1 v + A_2 u + b \in S, \qquad v \in T.$

Proof: Now problems I and I' have the same constraint set, therefore (u_0, v_0) is also feasible for problem I'. Suppose that for another feasible solution (u_1, v_1)

$$F(u_1, v_1) - F(u_0, v_0) < 0$$

i. e.

$$\operatorname{Re}(u_0^* Du_1 + v_0^* C v_1 + p^* v_1 - u_0^* Du_0 - v_0^* C v_0 - p^* v_0) < 0 \Rightarrow \operatorname{Re}[u_0^* D(u_1 - u_0) + v_0^* C (v_1 - v_0) + p^* (v_1 - v_0)] < 0. \tag{2.6}$$

Now from the fact that (u_0, v_0) and (u_1, v_1) are feasible solutions of problem I, it follows that (u_2, v_2) where

$$u_2 = u_0 + \lambda (u_1 - u_0)$$

$$v_2 = v_0 + \lambda (v_1 - v_0) \qquad 0 < \lambda < 1$$

is also feasible. But

$$\begin{aligned} f(u_2, v_2) - f(u_0, v_0) = \operatorname{Re}\Big[\frac{1}{2} [u_0 + \lambda (u_1 - u_0)]^* \times \\ \times D [u_0 + \lambda (u_1 - u_0)] + \frac{1}{2} [v_0 + \lambda (v_1 - v_0)]^* \times \\ \times C [v_0 + \lambda (v_1 - v_0)] + p^* [v_0 + \lambda (v_1 - v_0)] - \\ - \frac{1}{2} u_0^* \Big(Du_0 - \frac{1}{2} v_0^* C v_0 - p^* v_0\Big)\Big] \\ = \operatorname{Re}\Big[\lambda \{u_0^* D (u_1 - u_0) + v_0^* C (v_1 - v_0) + p^* (v_1 - v_0)\} + \\ + \frac{\lambda^2}{2} \{(u_1 - u_0)^* D (u_1 - u_0) + (v_1 - v_0)^* C (v_1 - v_0)\}\Big]. \end{aligned} \tag{2.7}$$

Right hand side of (2.7) can be made negative by taking λ sufficiently small and using (2.6). This contradicts the hypothesis that (u_0, v_0) is a minimizing solution of problem I. Hence (u_0, v_0) is minimizing solution of problem I'.

3. Duality theorems

Theorem 3.1: *If (u, v) and (ξ, η, ω) are feasible for Problems I and II respectively, then*

$$f(u, v) \geqq g(\xi, \eta, \omega).$$

Proof: From (2.2) and (2.3) we have

$$\operatorname{Re}\{v^* (-A_1^* \eta + C\,\xi + p)\} \geqq 0. \tag{3.1}$$

Similarly from (2.1) and (2.4) we have

$$\operatorname{Re}[(A_1 v + A_2 u + b)^* \eta] \geqq 0. \tag{3.2}$$

From (3.1) and (3.2) we can get

$$\operatorname{Re}[v^* C\,\xi + v^* p + u^* A_2^* \eta + b^* \eta] \geqq 0.$$

Now on using (2.5) we have

$$\operatorname{Re}[v^* C\,\xi + v^* p + u^* D\omega + b^* \eta] \geqq 0. \tag{3.3}$$

Since C and D are positive semidefinite hermitian matrices. Application of Lemma I in (3.3) shows that

$$\operatorname{Re}\left[\frac{1}{2} v^* C v + \frac{1}{2} \xi^* C \xi + v^* p + \frac{1}{2} u^* D u + \right.$$
$$\left. + \frac{1}{2} \omega^* D \omega + b^* \eta\right] \geqq 0 ,$$

or

$$\operatorname{Re}\left[\frac{1}{2} u^* D u + \frac{1}{2} v^* C v + p^* v\right] \geqq$$
$$\geqq \operatorname{Re}\left[-\frac{1}{2} \omega^* D \omega - \frac{1}{2} \xi^* C \xi - b^* \eta\right].$$

Hence $f(u, v) \geqq g(\xi, \eta, \omega)$.

Theorem 3.2: *If (u_0, v_0) is an optimal solution of problem I, then there exists an optimal solution $(\xi_0, \eta_0, \omega_0)$ of problem II such that $\xi_0 = v_0$ and $\omega_0 = u_0$ and $f(u_0, v_0) = g(\xi_0, \eta_0, \omega_0)$.*

Proof: Assume (u_0, v_0) is an optimal solution for Problem I, Lemma 2 shows that (u_0, v_0) solves problem I'.

Now dual to problem I' can be written as [5].

Problem II': Maximize $\operatorname{Re}(-b^* \eta)$,

$$\text{subject to } -A_1^* \eta + C v_0 + p \in T^* , \tag{3.4}$$
$$\eta \in S^* , \tag{3.5}$$
$$A_2^* \eta = D u_0 , \tag{3.6}$$

and there exists η_0 satisfying the constraints of problem II'. Also

$$\operatorname{Re}(-b^* \eta_0) = \operatorname{Re}(u_0^* D u_0 + v_0^* C v_0 + p^* v_0) . \tag{3.7}$$

We observe that $(\xi_0, \eta_0, \omega_0) = (v_0, \eta_0, \omega_0)$ is feasible for problem II. Also using (3.7) we get

$$g(v_0, \eta_0, \omega_0) = \operatorname{Re}\left(-\frac{1}{2} u_0^* D u_0 - \frac{1}{2} v_0^* C v_0 - b^* \eta_0\right)$$
$$= \operatorname{Re}\left(-\frac{1}{2} u_0^* D u_0 - \frac{1}{2} v_0^* C v_0 + \right.$$
$$\left. + u_0^* D u_0 + v_0^* C v_0 + p^* v_0\right)$$
$$= \operatorname{Re}\left(\frac{1}{2} u_0^* D u_0 + \frac{1}{2} v_0^* C v_0 + p^* v_0\right)$$
$$= f(u_0, v_0) . \tag{3.8}$$

Also from theorem (3.1) and equality (3.8) we have

$$f(u_0, v_0) \geqq g(\xi, \eta, \omega) , \quad \text{or} \quad g(\xi_0, \eta_0, \omega_0) \geqq g(\xi, \eta, \omega)$$

showing that (v_0, η_0, ω_0) is optimal solution for problem II.

Theorem 3.3: *If there exists a solution $(\xi_0, \eta_0, \omega_0)$ which maximizes problem II, then there exists a vector (u_0, v_0) which minimizes problem I and the extreme values of the two objective functions are equal.*

Proof: Assume that $(\xi_0, \eta_0, \omega_0)$ is an optimal solution for the dual problem.

Now writing the dual problem as a minimization problem. Minimize

$$- g(\xi, \eta, \omega) = \operatorname{Re}\left(\frac{1}{2} \omega^* D \omega + \frac{1}{2} \xi^* C \xi + b^* \eta\right),$$

subject to

$$- A_1^* \eta + C \xi + p \in T^* , \quad \eta \in S^* , \quad A_2^* \eta - D \omega = 0 ,$$

which can be rewritten as
Minimize

$$- g(\xi, \eta, \omega) = \operatorname{Re}\left\{\frac{1}{2} \omega^* D \omega + \frac{1}{2} \begin{pmatrix} \xi \\ \eta \end{pmatrix}^* \begin{pmatrix} C & 0 \\ 0 & 0 \end{pmatrix} \begin{pmatrix} \xi \\ \eta \end{pmatrix} + \right.$$
$$\left. + \begin{pmatrix} 0 \\ b \end{pmatrix}^* \begin{pmatrix} \xi \\ \eta \end{pmatrix}\right\},$$

subject to

$$\begin{bmatrix} C & -A_1^* \\ 0 & -A_2^* \end{bmatrix} \begin{bmatrix} \xi \\ \eta \end{bmatrix} + \begin{bmatrix} 0 \\ D \end{bmatrix} \omega + \begin{bmatrix} p \\ 0 \end{bmatrix} \in \begin{bmatrix} T^* \\ (C^n)^* \end{bmatrix},$$
$$\begin{bmatrix} \xi \\ \eta \end{bmatrix} \in \begin{pmatrix} C^n \\ S^* \end{pmatrix}.$$

This is now in the form of the problem I which has the following dual:

Maximize $f(y, x_1, x_2, v, u) =$

$$= \operatorname{Re}\left[-\frac{1}{2} y^* D y - \frac{1}{2} \begin{pmatrix} x_1 \\ x_2 \end{pmatrix}^* \begin{pmatrix} C & 0 \\ 0 & 0 \end{pmatrix} \begin{pmatrix} x_1 \\ x_2 \end{pmatrix} - \begin{pmatrix} p \\ 0 \end{pmatrix}^* \begin{pmatrix} v \\ u \end{pmatrix}\right],$$

subject to

$$\begin{bmatrix} C & 0 \\ 0 & 0 \end{bmatrix}^* \begin{bmatrix} x_1 \\ x_2 \end{bmatrix} - \begin{bmatrix} C^* & 0 \\ -A_1 & -A_2 \end{bmatrix} \begin{bmatrix} v \\ u \end{bmatrix} + \begin{bmatrix} 0 \\ b \end{bmatrix} \in \begin{pmatrix} C^n \\ S^* \end{pmatrix}^*,$$
$$\begin{bmatrix} v \\ u \end{bmatrix} \in \begin{bmatrix} T^* \\ (C^n)^* \end{bmatrix}^*, \qquad \begin{pmatrix} 0 \\ D \end{pmatrix}^* \begin{pmatrix} v \\ u \end{pmatrix} - D y = 0 .$$

Since S and T are closed convex cones, therefore, from lemma (3.4) of [3], it follows that $S = (S^*)^*$, $T = (T^*)^*$.

Hence $(C^n)^* = (0)$ and $(S^*)^* = S$. The above problem can now be written as
Maximize

$$f(y, x_1, x_2, v, u) = \operatorname{Re}\left(-\frac{1}{2} y^* D y - \frac{1}{2} x_1^* C x_1 - p^* v\right),$$

subject to

$$A_1 v + A_2 u + b \in S , \quad C^* x_1 = C^* v ,$$
$$v \in T , \quad D^* u - D y = 0 ,$$

which can be rewritten as
Minimize

$$f(u, v) = \operatorname{Re}\left(\frac{1}{2} u^* D u + \frac{1}{2} v^* C v + p^* v\right),$$

subject to

$$A_1 v + A_2 u + b \in S , \qquad v \in T .$$

Which is precisely the problem I.

Extreme values of the objective function are equal from theorem 3.2.

4. Special Cases

4.1. If we put $p = c$, $b = -b$, $A_2 = D$, $A_1 = A$, $C = B$, the dual problems reduce to the following pair.

Problem I: Minimize

$$f(u, v) = \operatorname{Re}\left(\frac{1}{2} u^* D u + \frac{1}{2} v^* B v + c^* v\right),$$

subject to

$$A v + D \omega - b \in S , \qquad v \in T .$$

Problem II: Maximize

$$g(\xi, \eta, \omega) = \operatorname{Re}\left(-\frac{1}{2} \omega^* D \omega - \frac{1}{2} \xi^* B \xi + b^* \eta\right),$$

s. t.

$$- A^* \eta + B \xi + c \in T^* , \qquad \eta \in S^* , \qquad D^* \eta - D \omega = 0 .$$

Problem II can be rewritten as:
Maximize

$$g(\xi, \eta, \omega) = \operatorname{Re}\left(-\frac{1}{2} \eta^* D \eta - \frac{1}{2} \xi^* B \xi + b^* \eta\right),$$

s. t.

$$- A^* \eta + B \xi + c \in T^* , \qquad \eta \in S^* ,$$

which are the dual problems considered by Abrams and Israel [5].

4.2. If $A_2 = D$ and we choose cones S and T as

$$S = \{x \in C^m: |\arg x| \leqq \alpha\},$$
$$T = \{z \in C^n: |\arg z| \leqq \beta\},$$

then for given $\alpha \in R_+^m$, $\beta \in R_+^n$, $\alpha \leqq \frac{\pi}{2}$, $\beta \leqq \frac{\pi}{2}$, the polars S^* and T^* are easily [6] seen to be

$$S^* = \left\{z \in C^m: |\arg z| \leqq \frac{\pi}{2} - \alpha\right\},$$

$$T^* = \left\{\omega \in C^n: |\arg \omega| \leqq \frac{\pi}{2} - \beta\right\};$$

then our primal and dual problems yield the following dual pair.

(i) Minimize

$$f(u, v) = \text{Re}\left(\frac{1}{2} u^* Du + \frac{1}{2} v^* C v + p^* v\right),$$

subject to

$$|\arg (A_1 v + Du + b)| \leqq \alpha, \qquad |\arg v| \leqq \beta,$$

(ii) Maximize

$$g(\xi, \eta, \omega) = \text{Re}\left[-\frac{1}{2} w^* Dw - \frac{1}{2} \xi^* C \xi - b^* \eta\right],$$

subject to

$$|\arg (A_1^* \eta + C \xi + p)| \leqq \frac{\pi}{2} - \beta,$$

$$|\arg \eta| \leqq \frac{\pi}{2} - \alpha, \qquad D^* \eta - D\omega = 0.$$

(ii) can be rewritten as

Maximize

$$g(\xi, \eta) = \text{Re}\left[-\frac{1}{2} \eta^* D\eta - \frac{1}{2} \xi^* C \xi - b^* \eta\right],$$

subject to

$$|\arg (-A_1^* \eta + C \xi + p)| \leqq \frac{\pi}{2} - \beta$$

$$|\arg \eta| \leqq \frac{\pi}{2} - \alpha.$$

Which are the dual problems considered by SHARMA and KAUL [3].

Acknowledgement

I am highly indebted to Dr. R. N. KAUL (Reader in Mathematics, Faculty of Mathematics, University of Delhi) for his continuous help and inspiration throughout the preparation of this note.

References

1 LEVINSON, N., Linear programming in complex space, Journal of Mathematical Analysis and Applications **14**, No. 1 (1966).
2 MOND, B., and HANSON, M. A., Symmetric duality for quadratic programming in complex space, Journal of Mathematical Analysis and Applications, **23**, No. 2 (1968).
3 SHARMA, S., KAUL, R. N., Symmetric and Self dual quadratic programs, Opsearch **6**, No. 3 (1969).
4 RANI, O., Duality for quadratic programming, ZAMM **52**, 314–316 (1972).
5 ABRAMS, R. A. and BEN ISRAEL, A., A duality theorem for complex quadratic programming. Journal of Optimization Theory and Applications, **4**, No. 4, Oct. (1969).
6 MOND, B., Non-linear, Non differentiable programming in complex space, Non-linear Programming by Mangasarian, Ritter & Rosen.

Eingereicht am 4. 11. 1972

Anschrift: Mrs. DAVINDER BHATIA, S. G. T. B. Khalsa College, University of Delhi, Delhi-7, India.

ZAMM **54**, 57–58 (1974)

S. L. PASSMAN

General Solutions for Motions with Convected Stretching

1. Introduction

In a previous work [1], the explicit form of the right CAUCHY-GREEN tensor for all motions in which the stretching tensor is convected but not diffused was obtained. Herein, I explicitly exhibit all motions, subject to a compatibility condition, with this property.

I employ the standard notations of modern continuum mechanics, as set forth by TRUESDELL and NOLL [2]. A short synopsis, including most of the identities utilized in this work, is given in [1].

2. Convection of stretching

I derive the general result for the transport of stretching for a particle. Consider the material derivative of $\boldsymbol{F}^T \boldsymbol{G} \boldsymbol{F}$. Since $\boldsymbol{G} = \dot{\boldsymbol{F}} \boldsymbol{F}^{-1}$, and $\boldsymbol{G}^* = \ddot{\boldsymbol{F}} \boldsymbol{F}^{-1}$

$$\frac{d}{dt}(\boldsymbol{F}^T \boldsymbol{G} \boldsymbol{F}) = \frac{d}{dt}(\boldsymbol{F}^T \dot{\boldsymbol{F}})$$
$$= \boldsymbol{F}^T \boldsymbol{G}^* \boldsymbol{F} + \boldsymbol{F}^T \boldsymbol{G}^T \boldsymbol{G} \boldsymbol{F}$$
$$= \boldsymbol{F}^T (\boldsymbol{G}^* + \boldsymbol{G}^T \boldsymbol{G}) \boldsymbol{F}. \tag{2.1}$$

Integrating along the path of a particle yields

$$\boldsymbol{G} = (\boldsymbol{F}^{-1})^T \left(\boldsymbol{G}_0 + \int_0^t \boldsymbol{F}^T (\boldsymbol{G}^* + \boldsymbol{G}^T \boldsymbol{G}) \boldsymbol{F}\, dt\right) \boldsymbol{F}^{-1}, \tag{2.2}$$

where I have taken the usual liberty of confusing the upper limit of the integral with the variable of integration. The first term, $(\boldsymbol{F}^{-1})^T \boldsymbol{G}_0 \boldsymbol{F}^{-1}$ on the right side of (2.2) is called the *convection* of $\boldsymbol{G}$, the second term the *diffusion* of $\boldsymbol{G}$. The skew part of (2.2) yields the equation of TRUESDELL [3]

$$\boldsymbol{W} = (\boldsymbol{F}^{-1})^T \left(\boldsymbol{W}_0 + \int_0^t \frac{1}{2} \boldsymbol{F}^T (\boldsymbol{G}^* - \boldsymbol{G}^{*T}) \boldsymbol{F}\, dt\right) \boldsymbol{F}^{-1}, \tag{2.3}$$

and the symmetric part yields the equation of CARSTOIU [4]

$$\boldsymbol{D} = (\boldsymbol{F}^{-1})^T \left(\boldsymbol{D}_0 + \int_0^t \boldsymbol{F}^T (\boldsymbol{D}^* + \boldsymbol{G}^T \boldsymbol{G}) \boldsymbol{F}\, dt\right) \boldsymbol{F}^{-1}, \tag{2.4}$$

where $\boldsymbol{G}_0 = \boldsymbol{D}_0 + \boldsymbol{W}_0$. Alternatively,

$$\overset{1}{\boldsymbol{A}} = (\boldsymbol{F}^{-1})^T \left(\overset{1}{\boldsymbol{A}}_0 + \int_0^t \boldsymbol{F}^T \overset{2}{\boldsymbol{A}} \boldsymbol{F}\, dt\right) \boldsymbol{F}^{-1}, \tag{2.5}$$

where $\overset{1}{\boldsymbol{A}}$ and $\overset{2}{\boldsymbol{A}}$ are the first and second RIVLIN-ERICKSEN tensors.

It is seen that the diffusion of $\boldsymbol{W}$ vanishes if and only if $\boldsymbol{G}^* = \boldsymbol{G}^{*T}$, a necessary and sufficient condition for circulation-preserving motion. PASSMAN and MARRIS [1] have shown that the diffusion of $\boldsymbol{D}$ vanishes if and only if

$$\boldsymbol{C} \equiv \boldsymbol{F}^T \boldsymbol{F} = 2\, \boldsymbol{D}_0 t + \boldsymbol{I}, \tag{2.6}$$

where $\boldsymbol{D}_0$ is positive definite and symmetric. Such a motion is called a *motion with convected stretching*. For such a motion

$$\boldsymbol{D} = (\boldsymbol{F}^{-1})^T \boldsymbol{D}_0 \boldsymbol{F}^{-1}. \tag{2.7}$$

I proceed to investigate this case.

3. Motions with convected stretching

By the polar decomposition theorem, $\boldsymbol{F} = \boldsymbol{R}\, \boldsymbol{U}$, and (2.6) becomes

$$\boldsymbol{U}^2 = 2\, \boldsymbol{D}_0 t + \boldsymbol{I}. \tag{3.1}$$

Since $\boldsymbol{U}^2$ is positive definite, there is a unique square root,

$$\boldsymbol{U} = \sqrt{2\, \boldsymbol{D}_0 t + \boldsymbol{I}}. \tag{3.2}$$

Let $\boldsymbol{R}(\boldsymbol{X}, t)$ be an orthogonal tensor satisfying $\boldsymbol{R}(\boldsymbol{X}, 0) = \boldsymbol{I}$, such that

$$\nabla_{\boldsymbol{X}} \boldsymbol{R}\, \boldsymbol{U} = (\nabla_{\boldsymbol{X}} \boldsymbol{R}\, \boldsymbol{U})^T. [1] \tag{3.3}$$

[1]) If x is expressed in terms of coordinates x^i and X is expressed in terms of Cartesian coordinates X^I, a component form of this equation is $\frac{\partial}{\partial X^J}(R_{IK}\, U^{Ki}) = \frac{\partial}{\partial X^I}(R_{JK}\, U^{Ki})$.

One choice of $\boldsymbol{R}(\boldsymbol{X}, t)$ satisfying these conditions is $\boldsymbol{R}(\boldsymbol{X}, t) \equiv \boldsymbol{I}$. Then all motions with convected stretching have a deformation gradient of the form

$$\boldsymbol{F} = \boldsymbol{R} \sqrt{2 \boldsymbol{D}_0 t + \boldsymbol{I}}. \tag{3.4}$$

Let $\boldsymbol{X}_0$ be an arbitrary fixed value of $\boldsymbol{X}$. The solution of (3.4) which satisfies the correct initial condition is

$$\chi(\boldsymbol{X}, t) = \boldsymbol{X} + \int_{\boldsymbol{X}_0}^{\boldsymbol{X}} [\boldsymbol{R}(\boldsymbol{X}, t) \sqrt{2 \boldsymbol{D}_0(\boldsymbol{X}) t + \boldsymbol{I}}]\, d\boldsymbol{X}. \tag{3.5}$$

Collecting hypotheses, I obtain the following result. *For every motion with convected stretching, there exists a positive definite symmetric tensor function $\boldsymbol{D}_0(\cdot)$, an orthogonal tensor function $\boldsymbol{R}(\cdot, \cdot)$ and a fixed point $\boldsymbol{X}_0$ such that the motion may be written in the form* (3.5).

As a type of converse: *For every positive definite symmetric tensor function $\boldsymbol{D}_0(\cdot)$ and every orthogonal tensor function $\boldsymbol{R}(\cdot, \cdot)$ such that $\sqrt{2 \boldsymbol{D}_0(\boldsymbol{X}) t + \boldsymbol{I}}$ exists and* (3.3) *is satisfied, there exists a motion with convected stretching which is given by the explicit formula* (3.5).

4. Some homogeneous motions with convected stretching

In (3.5), choose

$$\boldsymbol{R}(\boldsymbol{X}, t) = \boldsymbol{I}, \quad \boldsymbol{D}_0(\boldsymbol{X}) = \boldsymbol{D}_c \text{ (a constant)}, \quad \boldsymbol{X}_0 = 0. \tag{4.1}$$

In this case

$$\chi(\boldsymbol{X}, t) = \sqrt{2 \boldsymbol{D}_c t + \boldsymbol{I}} \cdot \boldsymbol{X} \tag{4.2}$$

so that the particle paths are straight lines. The deformation gradient is given by

$$\boldsymbol{F}(\boldsymbol{X}, t) = \sqrt{2 \boldsymbol{D}_c t + \boldsymbol{I}}, \tag{4.3}$$

so that $\sqrt{2 \boldsymbol{D}_c t + \boldsymbol{I}}$ exists, (4.2) is a special type of homogeneous deformation, $\boldsymbol{F}$ is independent of $\boldsymbol{X}$, and (4.3) is satisfied. Thus, (4.2) is a motion with convected stretching. Since $\boldsymbol{D}_c$ is symmetric, it has real principal axes. Referred to these axes, it has the form

$$[\boldsymbol{D}_c] = \operatorname{diag}(D_1, D_2, D_3). \tag{4.4}$$

Assume D_1, D_2 and D_3 are non-negative. The $\boldsymbol{F}$ and $\boldsymbol{F}^T$ are given by

$$[\boldsymbol{F}] = [\boldsymbol{F}^T] = \operatorname{diag}(\sqrt{2 D_1 t + 1}, \sqrt{2 D_2 t + 1}, \sqrt{2 D_3 t + 1}) \tag{4.5}$$

while $\boldsymbol{F}^{-1}$ and $(\boldsymbol{F}^{-1})^T$ are

$$[\boldsymbol{F}^{-1}] = [(\boldsymbol{F}^{-1})]^T = \operatorname{diag}\left(\frac{1}{\sqrt{2 D_1 t + 1}}, \frac{1}{\sqrt{2 D_2 t + 1}}, \frac{1}{\sqrt{2 D_3 t + 1}}\right), \tag{4.6}$$

also

$$[\dot{\boldsymbol{F}}] = \operatorname{diag}\left(\frac{D_1}{\sqrt{2 D_1 t + 1}}, \frac{D_2}{\sqrt{2 D_2 t + 1}}, \frac{D_3}{\sqrt{2 D_3 t + 1}}\right). \tag{4.7}$$

Thus,

$$[\boldsymbol{G}] = [\boldsymbol{D}] = \operatorname{diag}\left(\frac{D_1}{\sqrt{2 D_1 t + 1}}, \frac{D_2}{\sqrt{2 D_2 t + 1}}, \frac{D_3}{\sqrt{2 D_3 t + 1}}\right). \tag{4.8}$$

The right Cauchy-Green tensor is

$$[\boldsymbol{C}] = \operatorname{diag}(2 D_1 t + 1, 2 D_2 t + 1, 2 D_3 t + 1), \tag{4.9}$$

and the relation (2.7) becomes

$$\operatorname{diag}\left(\frac{D_1}{2 D_1 t + 1}, \frac{D_2}{2 D_2 t + 1}, \frac{D_3}{2 D_3 t + 1}\right) = \left[\operatorname{diag}\left(\frac{1}{\sqrt{2 D_1 t + 1}}, \frac{1}{\sqrt{2 D_2 t + 1}}, \frac{1}{\sqrt{2 D_3 t + 1}}\right)\right] \times [\operatorname{diag}(D_1, D_2, D_3)] \left[\operatorname{diag}\left(\frac{1}{\sqrt{2 D_1 t + 1}}, \frac{1}{\sqrt{2 D_2 t + 1}}, \frac{1}{\sqrt{2 D_3 t + 1}}\right)\right]. \tag{4.10}$$

Let $\boldsymbol{x}$ and $\boldsymbol{X}$ both be expressed in terms of rectangular Cartesian coordinates. The relation (4.2) becomes

$$x^1 = \sqrt{2 D_1 t + 1}\, X^1, \quad x^2 = \sqrt{2 D_2 t + 1}\, X^2, \quad x_3 = \sqrt{2 D_3 t + 1}\, X^3. \tag{4.11}$$

The velocity field, in terms of spatial coordinates, is given by

$$\dot{x}_1 = \frac{D_1}{2 D_1 t + 1} x^1, \quad \dot{x}_2 = \frac{D_2}{2 D_2 t + 1} x_2, \quad \dot{x}_3 = \frac{D_3}{2 D_3 t + 1} x^3. \tag{4.12}$$

The streamlines, solutions of

$$\frac{dx^1}{\dot{x}^1} = \frac{dx^2}{\dot{x}^2} = \frac{dx^3}{\dot{x}^3}, \tag{4.13}$$

are given, for D_1, D_2, and D_3 non-vanishing, by

$$(x^1)^{(2 D_1 t + 1)/D_1} = (K_2 x^2)^{(2 D_2 t + 1)/D_2} = (K_3 x^3)^{(2 D_3 t + 1)/D_3}. \tag{4.14}$$

Thus, although the particle paths are straight lines, the streamlines are of a complicated nature.

Two special cases are of particular interest:

a) Unsteady Expansion. This is the case

$$D_1 = D_2 = D_3. \tag{4.15}$$

Let D denote the common value of these three quantities, and assume $D > 0$. The motion (4.2) besomes

$$\boldsymbol{x} = \sqrt{2 D t + 1}\, \boldsymbol{X} \tag{4.16}$$

and all of the streamlines are straight. However, it is not in general true that for all time the streamline through a point and the particle path through that point coincide.

b) Unsteady Simple Extension. Here, for example,

$$D_1 + D_2 = D_3 \tag{4.17}$$

with all three constants positive. The motion (4.2) becomes

$$x^1 = \sqrt{2 D_1 t + 1}\, X^1, x^2 = \sqrt{2 D_2 t + 1}\, X^2, x^3 = \sqrt{2 D_2 t + 1}\, X^3. \tag{4.18}$$

In this case, the streamlines in the planes x^1 = constant are straight. In general, however, the streamlines exhibit a complicated behavior.

References

1 Passman, S. L., and Marris, A. W., Convection of stretching, ZAMM **53**, 127–130 (1973).
2 Truesdell, C., and Noll, W., The Non-Linear Field Theories of Mechanics, Handbuch der Physik III/3, Berlin 1965.
3 Truesdell, C., Généralisation de la formule de Cauchy et des théorèmes de Helmholtz au mouvement d'un milieu quelconque, Comptes Rendus de l'Academie des Sciences, Paris **227**, 757–759 (1948).
4 Carstoiu, I., Vorticity and deformation in fluid mechanics, Journal of Rational Mechanics and Analysis **3**, 691–712 (1954).

Eingereicht am 17. 1. 1973

Anschrift: S. L. Passman, Georgia Institute of Technology, Atlanta, Georgia 30332, USA

ZAMM **54**, 58–62 (1974)

V. Scharf

Optimuminvariante Bereiche der Zielfunktionskoeffizienten linearer Programme

Gegeben sei ein lineares Programm in der Form

$$\begin{array}{lll} \text{Min!}\ Z(c, x) = c'\, x, & c \in \mathbb{R}^n,\ x \in \mathbb{R}^n, & \\ A\, x = b & A = (a_{ij}),\quad b \in \mathbb{R}^m & (1) \\ x \geqq \theta_n & i = 1\,(1)\,m,\quad j = 1\,(1)\,n. & \end{array}$$

Ist x_0 eine Optimallösung dieses Problems, so nennen wir einen *Störungsvektor* $\Delta c \in \mathbb{R}^n$ *optimuminvariant* (abgekürzt Δc = oiv), wenn x_0 auch optimal ist bezüglich der Zielfunktion

$$Z(c + \Delta c, x) = (c + \Delta c)'\, x.$$

Die Kenntnis gewisser Bereiche optimuminvarianter Störungen der Zielfunktionskoeffizienten ist für die Praxis oft sehr nützlich, besonders dann, wenn mit relativ unsicheren Daten gerechnet werden muß. Deshalb enthalten moderne Programmsysteme zur linearen Optimierung, wie z. B. das IBM-MPS/360, auch Programme, die im Rahmen der Sensitivitätsanalyse über den Gültigkeitsbereich der Lösung Aussagen machen. Die diesen Programmen zugrunde liegenden Verfahren sind z. B. in [1], [2], [3] ausführlich beschrieben. Die danach zu ermittelnden optimuminvarianten Störungen sind in der Regel von der Form

$$\Delta c = \Delta c_i \, e_i = \begin{pmatrix} 0 \\ \vdots \\ \Delta c_i \\ \vdots \\ 0 \end{pmatrix}, \quad e_i = i\text{-ter Einheitsvektor,}$$

betreffen also nur einen Zielfunktionskoeffizienten.

Gegenstand dieser Untersuchung ist daher die Frage nach höherdimensionalen Bereichen optimuminvarianter Störungen Δc. Dabei soll gezeigt werden, daß man sowohl aus den Daten des Optimaltableaus als auch aus den *eindimensionalen* Schwankungsbereichen, wie sie als Ergebnis einer Empfindlichkeitsanalyse anfallen, relativ leicht Aussagen über Bereiche simultaner optimuminvarianter Störungen mehrerer Zielfunktionskoeffizienten gewinnen kann.

Um die folgenden Darstellungen möglichst zu vereinfachen, sei angenommen, daß in der Optimallösung die ersten m Variablen $x_1, \ldots, x_m$ basisch sind.

Bezeichnet man mit $\bar{A}, \bar{b}, \bar{c}$ die Daten des Optimaltableaus, so lautet die Lösung:

$$\begin{aligned} x_k &= \bar{b}_k , & k &= 1(1)\, m \\ x_j &= 0 , & j &= m + 1(1)\, n . \end{aligned} \tag{2}$$

Der Zusammenhang zwischen Ausgangs- und Optimaltableau läßt sich leicht mittels Blockmatrizen darstellen:

$$\left(\begin{array}{ccc|c|c} & \bar{A} & & 0 & \bar{b}_1 \\ & & & \vdots & \vdots \\ & & & 0 & \bar{b}_m \\ \hline \bar{c}_1 & \cdots & \bar{c}_n & 1 & -Z_0 \end{array}\right) = \left(\begin{array}{ccc|c} & A_{mm} & & 0 \\ & & & \vdots \\ & & & 0 \\ \hline c_1 & \cdots & c_m & 1 \end{array}\right)^{-1} \cdot \left(\begin{array}{ccc|c|c} & A & & 0 & b_1 \\ & & & \vdots & \vdots \\ & & & 0 & b_m \\ \hline c_1 & \cdots & c_n & 1 & 0 \end{array}\right).$$

Dabei ist $A_{mm} = (a_{ik})$, $i = 1(1)\, m$, $k = 1(1)\, m$ und

$$\left(\begin{array}{ccc|c} & A_{mm} & & 0 \\ & & & \vdots \\ & & & 0 \\ \hline c_1 & \cdots & c_m & 1 \end{array}\right)^{-1} = \left(\begin{array}{ccc|c} & A_{mm}^{-1} & & 0 \\ & & & \vdots \\ & & & 0 \\ \hline d_1 & \cdots & d_m & 1 \end{array}\right),$$

$$(d_1, \ldots, d_m) = -\,(c_1, \ldots, c_m)\, A_{mm}^{-1} .$$

Damit gilt zwischen c und $\bar{c}$ die Beziehung

$$\bar{c}' = -\,(c_1, \ldots, c_m)\, A_{mm}^{-1} A + c' \qquad \text{bzw.}$$

$$\bar{c} = c - \bar{A}' \begin{pmatrix} c_1 \\ \vdots \\ c_m \end{pmatrix}. \tag{3}$$

Nach dem Optimalitätskriterium ist die durch (2) gegebene Basislösung genau dann optimal, wenn $\bar{c} \geqq \theta$ ist.

Daraus folgt unmittelbar der für die folgenden Betrachtungen grundlegende

Satz 1: $\Delta c = \text{oiv} \Leftrightarrow$

$$\bar{A}' \begin{pmatrix} \Delta c_1 \\ \vdots \\ \Delta c_m \end{pmatrix} \leqq \bar{c} + \Delta c \qquad \textit{bzw.} \tag{4}$$

$$\sum_{i=1}^{m} \bar{a}_{ki}\, \Delta c_k \leqq \bar{c}_i + \Delta c_i , \qquad i = m + 1(1)\, n . \tag{4a}$$

Aus (4) bzw. (4a) erkennt man, daß die optimuminvarianten Δc eine konvexe Teilmenge des $\mathbb{R}^n$ bilden, und zwar ein Simplex, das auch Punkte mit $\Delta c \ngeq \theta$ enthalten kann.

Die nachstehend behandelten Fälle werden jeweils solche Teilbereiche dieses Simplex erfassen, die sich aus den Daten des Optimaltableaus leicht bestimmen lassen.

Fall 1:

$\Delta c_1 = \cdots = \Delta c_m = 0$, es werden also nur Störungen der zu Nichtbasisvariablen gehörigen Zielfunktionskoeffizienten, wir wollen sie kurz *nichtbasische Koeffizienten* nennen, zugelassen. Es folgt

$$\Delta c' = (0, \ldots, 0, \Delta c_{m+1}, \ldots, \Delta c_n) = \text{oiv} \leftrightarrow \Delta c_i \geqq -\bar{c}_i , \quad i = m + 1(1)\, n . \tag{5}$$

Fall 2:

$$\Delta c = \Delta c_k\, e_k , \quad 1 \leqq k \leqq m .$$

Dies ist der schon erwähnte Fall einer *eindimensionalen* Störung, d. h. nur ein basischer Koeffizient darf variiert werden, während alle übrigen Koeffizienten unverändert bleiben.

Es gilt:

$$\Delta c = \Delta c_k\, e_k = \text{oiv} \leftrightarrow$$

$$d_k = \operatorname*{Max}_{\substack{\bar{a}_{ki} < 0 \\ i = m+1(1)\, n}} \left\{ \frac{\bar{c}_i}{\bar{a}_{ki}} \right\} \leqq \Delta c_k \leqq \operatorname*{Min}_{\substack{\bar{a}_{ki} > 0 \\ i = m+1(1)\, n}} \left\{ \frac{\bar{c}_i}{\bar{a}_{ki}} \right\} = D_k . \tag{6}$$

Die durch die Formeln (5) u. (6) definierten Schranken werden z. B. im Rahmen der LPANALYSIS-Routine des Programmsystems IBM-LPS/360 ausgedruckt.

Geometrisch bedeuten die nach (6) bestimmten Schranken d_k und D_k die durch das oiv-Simplex (4) auf der k-ten Koordinatenachse des $(\Delta c_1, \ldots, \Delta c_n)$-Raumes gebildeten Achsenabschnitte.

Fall 3:

Sei

$$f(c_i) = \begin{cases} 0, & \Delta c_i \equiv 0, \text{ d. h. } c_i \text{ bleibt ungestört} \\ 1, & \Delta c_i \neq 0, \text{ d. h. } c_i \text{ darf gestört werden.} \end{cases}$$

$$\Delta c' = (f(c_1)\, \Delta c_1, \ldots, f(c_m)\, \Delta c_m, 0, \ldots, 0) .$$

Dieser Fall besagt also, daß einige oder alle basischen Koeffizienten verändert werden dürfen.

Gesucht ist eine maximale Zahl K, so daß $\Delta c = \text{oiv}$ für

$$|\Delta c_k| \leqq K, \quad 1 \leqq k \leqq m .$$

Aus (4a) folgt als notwendige und hinreichende Bedingung:

$$\sum_{k=1}^{m} \bar{a}_{ki}\, f(c_k)\, \Delta c_k \leqq \bar{c}_i , \qquad i = m + 1(1)\, n .$$

Dies ist für alle Δc_k mit $|\Delta c_k| \leqq K$ genau dann erfüllt, wenn

$$K \sum_{k=1}^{m} |\bar{a}_{ki}|\, f(c_k) \leqq \bar{c}_i \quad \text{für alle } i \text{ mit } m + 1 \leqq i \leqq n.$$

Wir erhalten also:

$$\Delta c' = (f(c_1)\, \Delta c_1, \ldots, f(c_m)\, \Delta c_m, 0, \ldots, 0) = \text{oiv} \quad \text{für}$$

$$|\Delta c_k| \leqq K = \operatorname*{Min}_{m+1 \leqq i \leqq n} \left\{ \frac{\bar{c}_i}{\sum_{k=1}^{m} |\bar{a}_{ki}|\, f(c_k)} \right\}. \tag{7}$$

Die Zahl K ist die halbe Kante des größten achsenparallelen Würfels mit Mittelpunkt im Koordinatenursprung, der sich in das oiv-Simplex im $(\Delta c_1, \ldots, \Delta c_m)$-Raum einbeschreiben läßt. Sind alle $f(c_k)$ für $k = 1(1)\, m$ gleich 1, so erhält man K dadurch, daß man jeden Koeffizienten $\bar{c}_i$ des Optimaltableaus durch die Summe der Beträge der i-ten Spalte der Matrix $\bar{A}$ dividiert und von diesen Quotienten das Minimum bildet.

Als Verallgemeinerung von (7) ergibt sich für den Fall

$$\Delta c' = (f(c_1)\, \Delta c_1, \ldots, f(c_n)\, \Delta c_n)$$

die Aussage

$$\Delta c = \text{oiv} \quad \text{für } |\Delta c_i| \leqq K$$

$$= \min_{m+1 \leqq i \leqq n} \left\{ \frac{\bar{c}_i}{\sum_{k=1}^{m} |\bar{a}_{ki}| f(c_k) + f(c_i)} \right\}. \tag{8}$$

Für $f(c_i) = 0$, $m + 1 \leqq i \leqq n$, erhält man wieder Formel (7), ist andererseits $f(c_k) = 0$ für $1 \leqq k \leqq m$, so ergibt sich $K = \min_{m+1 \leqq i \leqq n} |\bar{c}_i|$ als Spezialfall von Formel (5).

Fall 4:

Sei

$$f(c_i) = 0 \quad \text{für} \quad m + 1 \leqq i \leqq n\,,$$

gesucht sind Zahlen K_j, so daß

$$\Delta c' = (f(c_1)\,\Delta c_1, \ldots, f(c_m)\,\Delta c_m,\ 0, \ldots, 0) = \text{oiv}$$

für $|\Delta c_j| \leqq K_j$.

Es soll also statt eines Würfels ein Quader in das oiv-Simplex im $(\Delta c_1, \ldots, \Delta c_m)$-Raum einbeschrieben werden. Diese Aufgabe ist nicht eindeutig lösbar, denn es gelten jetzt für die K_j die Restriktionen:

$$\sum_{j=1}^{m} |\bar{a}_{ji}|\, f(c_j)\, K_j \leqq \bar{c}_i\,, \qquad m + 1 \leqq i \leqq n\,. \tag{9}$$

Da die Optimallösung x_0 auch optimal bezüglich der Zielfunktion $Z(c + \Delta c, x)$ sein soll, gilt:

$$Z(c + \Delta c, x_0) - Z(c, x_0) = \sum_{i=1}^{n} \Delta c_i \overset{0}{x}_i\,.$$

Für $|\Delta c_i| \leqq K_i$ folgt daraus

$$\max_{|\Delta c_i| \leqq K_i} |Z(c + \Delta c, x_0) - Z(c, x_0)| = \sum_{i=1}^{n} f(c_i)\, K_i \overset{0}{x}_i$$

$$= \sum_{j=1}^{m} f(c_j)\, K_j \bar{b}_j\,.$$

Das lineare Programm

$$\text{Max!} \sum_{j=1}^{m} f(c_j)\, \bar{b}_j\, K_j$$

$$\sum_{j=1}^{m} |\bar{a}_{ji}|\, f(c_j)\, K_j \leqq \bar{c}_i, \quad i = m + 1(1)\,n \tag{10}$$

$$K_j \geqq 0\,,$$

liefert demnach als Lösung einen achsenparallelen Quader mit Mittelpunkt im Koordinatenursprung, der ganz im oiv-Simplex enthalten ist und durch den der maximale, durch die zugelassenen Störungen bedingte Zielfunktionsbereich bei gleicher Optimallösung überdeckt wird.

Interessiert man sich für möglichst große Schwankungsbereiche solcher Koeffizienten c_j, die einen wesentlichen Anteil am Optimalwert der Zielfunktion liefern, so würde man z. B. die Zielfunktion

$$\sum_{j=1}^{m} (-\, c_j\, \bar{b}_j\, f(c_j))\, K_j$$

unter den Nebenbedingungen (9) maximieren, denn $c_j \bar{b}_j$ ist der durch die Basisvariable x_j gelieferte Anteil am optimalen Z-Wert.

Die bisher betrachteten Fälle setzen die Kenntnis der Daten des Optimaltableaus voraus, was z. B. bei der manuellen Durchrechnung kleinerer Programme keinerlei Einschränkung bedeutet. Bei der maschinellen Lösung größerer LP-Probleme unter Verwendung geeigneter Programmsysteme sind jedoch u. U. nicht sämtliche Daten des Optimaltableaus unmittelbar bekannt.

Es kann aber davon ausgegangen werden, daß z. B. die durch (5) und (6) beschriebenen Schranken für die Fälle 1 und 2 als Ergebnis einer Empfindlichkeitsanalyse verfügbar sind. Aus diesen Schranken lassen sich nun leicht und ohne wesentlichen zusätzlichen Aufwand weitere oiv-Bereiche der Zielfunktionskoeffizienten ableiten.

Die Punkte θ, $p_j = d_j\, e_j$, $P_j = D_j\, e_j$, $j = 1(1)\, m$

$$P_i = -\, \bar{c}_i\, e_i\,, \quad m + 1 \leqq i \leqq n\,,$$

sind ja im oiv-Simplex enthalten, also sind auch alle Elemente der konvexen Hülle dieser Punkte optimuminvariante Störungsvektoren.

Das durch diese konvexe Hülle gegebene Teilsimplex besteht aus den Punkten

$$\Delta c = \sum_{j=1}^{m} \lambda_j\, p_j + \sum_{j=1}^{m} \mu_j\, P_j + \sum_{i=m+1}^{n} \varkappa_i\, P_i\,, \quad \text{mit}$$
$$\sum_{j=1}^{m} (\lambda_j + \mu_j) + \sum_{i=m+1}^{n} \varkappa_i \leqq 1\,, \quad \lambda_j, \mu_j, \varkappa_i \geqq 0\,. \tag{11}$$

Aus (11) folgt für diese Δc:

$$\lambda_j\, d_j + \mu_j\, D_j = \Delta c_j\,, \qquad 1 \leqq j \leqq m$$

$$-\, \varkappa_i\, \bar{c}_i = \Delta c_i\,, \qquad m + 1 \leqq i \leqq n\,.$$

Es soll angenommen werden, daß die Werte d_j, D_j, $\bar{c}_i$ von Null verschieden sind.

Für ein bestimmtes Δc_j betrachten wir das lineare Programm

$$\text{Min!}\ z_j = \lambda_j + \mu_j\,, \quad \lambda_j\, d_j + \mu_j\, D_j = \Delta c_j,\ \lambda_j \geqq 0,\ \mu_j \geqq 0\,.$$

Als Lösung erhält man einen der beiden Achsenabschnitte der im λ_j, μ_j-Koordinatensystem dargestellten Geraden

$$d_j\, \lambda_j + D_j\, \mu_j = \Delta c_j\,.$$

Da d_j und D_j verschiedene Vorzeichen besitzen, so ist die Lösung dieses Programmes gleich dem positiven dieser beiden Achsenabschnitte, d. h. es ist $z_j = \frac{\Delta c_j}{d_j}$ für $\Delta c_j < 0$ und $z_j = \frac{\Delta c_j}{D_j}$ für $\Delta c_j \geqq 0$. Genau dann, wenn

$$\sum_{j=1}^{m} z_j - \sum_{i=m+1}^{n} \frac{\Delta c_i}{\bar{c}_i} \leqq 1$$

gilt, ist Δc aus dem durch (11) beschriebenen Simplex. Daraus ergibt sich der

Satz 2: *Gilt*

$$\sum_{\substack{j=1 \\ \Delta c_j < 0}}^{m} \frac{\Delta c_j}{d_j} + \sum_{\substack{j=1 \\ \Delta c_j \geqq 0}}^{m} \frac{\Delta c_j}{D_j} - \sum_{i=m+1}^{n} \frac{\Delta c_i}{\bar{c}_i} \leqq 1\,, \tag{12}$$

so ist $\Delta c =$ oiv.

Satz 2 und Formel (12) können z. B. dazu benutzt werden, die Optimuminvarianz einer gegebenen Störung Δc zu überprüfen, allerdings stellt dieser Satz nur eine hinreichende Bedingung dar. Sei

$$s_j \leqq \Delta c_j \leqq S_j\,, \qquad 1 \leqq j \leqq m$$

$$-\, s_i \leqq \Delta c_i\,, \qquad m + 1 \leqq i \leqq n\,.$$

Setzt man

$$s_j = \frac{d_j}{n}, \qquad S_j = \frac{D_j}{n}, \qquad s_i = \frac{\bar{c}_j}{n},$$

so gilt

$$\frac{\Delta c_j}{d_j} \leqq \frac{s_j}{d_j} = \frac{1}{n} \qquad \text{für } \Delta c_j < 0$$

$$\frac{\Delta c_j}{D_j} \leqq \frac{S_j}{D_j} = \frac{1}{n} \qquad \text{für } \Delta c_j \geqq 0$$

$$\frac{-\, \Delta c_i}{\bar{c}_i} \leqq \frac{s_i}{\bar{c}_i} = \frac{1}{n}\,.$$

Daraus folgt nach Satz 2, daß alle Δc, deren Komponenten in den Schranken $\left[\frac{d_j}{n}, \frac{D_j}{n}\right]$, $\left[\frac{-c_i}{n}, \infty\right]$ liegen, optimuminvariant sind.

Eine Verallgemeinerung dieser Aussage erhält man durch den folgenden

Satz 3: *Sind die Vektoren* $\Delta\overset{(0)}{c} + \lambda_i \Delta\overset{(i)}{c} = \text{oiv}$ *für*

$s_i \leqq \lambda_i \leqq S_i, \; i = 1(1)\,r,$

so sind die Vektoren

$$\Delta\overset{(0)}{c} + \sum_{i=1}^{r} \tilde\lambda_i \Delta\overset{(i)}{c} = \text{oiv},$$

wenn gilt

$$\mu_i s_i \leqq \tilde\lambda_i \leqq \mu_i S_i, \quad \mu_i \geqq 0, \quad \sum_{i=1}^{r} \mu_i = 1. \tag{13}$$

Da nämlich die konvexe Hülle der Punkte $\Delta\overset{(0)}{c} + \lambda_i \Delta\overset{(i)}{c}$ im oiv-Simplex enthalten ist, ist jeder Vektor

$$\sum_{i=1}^{r} \mu_i (\Delta\overset{(0)}{c} + \lambda_i \Delta\overset{(i)}{c}) = \text{oiv}$$

für $\sum_{i=1}^{r} \mu_i = 1$, $\mu_i \geqq 0$ und $s_i \leqq \lambda_i \leqq S_i$.

Daraus folgt

$$\Delta\overset{(0)}{c} + \sum_{i=1}^{r} \mu\, \lambda_i \Delta\overset{(i)}{c} = \Delta\overset{(0)}{c} + \sum_{i=1}^{r} \tilde\lambda_i \Delta\overset{(i)}{c} = \text{oiv}$$

für $\mu_i s_i \leqq \tilde\lambda_i \leqq \mu_i S_i$.

Dieser Satz bietet die Möglichkeit, auf mannigfache Weise oiv-Bereiche zu konstruieren.

Beispiel 1:

$$\Delta\overset{(0)}{c} = \theta, \quad \Delta\overset{(i)}{c} = e_i, \quad \mu_i = \frac{1}{r}, \quad i = 1(1)\,r.$$

Die Voraussetzungen des Satzes sind erfüllt für $d_i \leqq \lambda_i \leqq D_i$, wobei die Zahlen d_i und D_i nach (5) bzw. (6) zu ermitteln sind. Es folgt unmittelbar der in der Praxis besonders einfach zu bestimmende oiv-Bereich:

$$\frac{d_i}{r} \leqq \tilde\lambda_i \leqq \frac{D_i}{r}, \qquad i = 1(1)\,r. \tag{14a}$$

Für $\Delta\overset{(0)}{c} = (D_j/r)\, e_j$, $\Delta\overset{(i)}{c} = e_i$, sind die Voraussetzungen erfüllt, wenn $\frac{d_i}{r} \leqq \lambda_i \leqq \frac{D_i}{r}$, $i \neq j$

$$\frac{r\, d_j - D_j}{r} \leqq \lambda_j \leqq \frac{r-1}{r} D_j.$$

Man erhält den Bereich

$$\frac{d_i}{r^2} \leqq \tilde\lambda_i \leqq \frac{D_i}{r^2}, \qquad i \neq j$$

$$\frac{(r-1)\, D_j + r\, d_j}{r^2} \leqq \tilde\lambda_j \leqq \frac{(2\,r-1)\, D_j}{r^2}.$$

Für $\Delta\overset{(0)}{c} = \frac{d_j}{r} e_j$ ergibt sich entsprechend

$$\frac{d_i}{r^2} \leqq \tilde\lambda_i \leqq \frac{D_i}{r^2}, \qquad i \neq j$$

$$\frac{(2\,r-1)\, d_j}{r^2} \leqq \tilde\lambda_j \leqq \frac{r\, D_j + (r-1)\, d_j}{r^2}.$$

Da $r\, D_j + (r-1)\, d_j \geqq (r-1)\, D_j + r\, d_j$, erhält man als Vereinigungsmenge den durch (14b) beschriebenen Quader:

$$\frac{d_i}{r^2} \leqq \tilde\lambda_i \leqq \frac{D_i}{r^2}, \quad i \neq j$$

$$\frac{(2\,r-1)\, d_j}{r^2} \leqq \tilde\lambda_j \leqq \frac{(2\,r-1)\, D_j}{r^2}. \tag{14b}$$

Formeln (14a) und (14b) sind Spezialfälle der folgenden Formel (14):

$$\frac{d_i}{r^t} \leqq \tilde\lambda_i \leqq \frac{D_i}{r^t}, \qquad i \neq j$$

$$q_t\, d_j \leqq \tilde\lambda_j \leqq q_t\, D_j, \qquad t = 1, 2, \ldots \tag{14}$$

$$q_1 = \frac{1}{r}, \qquad q_{t+1} = \frac{(r-1)\, q_t + 1}{r}.$$

Die Folge q_t strebt für wachsende t monoton gegen 1. Es ergeben sich also Quader, die entlang der Δc_j-Achse wachsen, dafür aber in den übrigen Richtungen immer schmaler werden, bis im Grenzfall $t \to \infty$ der nach (6) definierte eindimensionale oiv-Bereich entsteht.

Beispiel 2:

Die Gewichte μ_i werden gemäß der *Bedeutung* der Variablen verteilt, nämlich nach der Formel

$$\mu_i = \frac{\bar b_i}{\sum_{j=1}^{r} \bar b_j}.$$

Die Koeffizienten derjenigen Variablen, die in der Optimallösung einen hohen Wert erreichen, werden nach Satz 3 ihrem Gewicht entsprechend möglichst breite Schwankungsbereiche erhalten. Setzt man wieder $\Delta\overset{(0)}{c} = \theta$ and $\Delta\overset{(i)}{c} = e_i$, so ergibt sich der oiv-Bereich

$$d_i \frac{\bar b_i}{\sum_{j=1}^{r} \bar b_j} \leqq \tilde\lambda_i \leqq D_i \frac{\bar b_i}{\sum_{j=1}^{r} \bar b_j}, \qquad i = 1(1)\,r. \tag{15}$$

Beispiel 3:

Bei der eindimensionalen parametrischen Optimierung macht man den Ansatz $\Delta c = t\, p$, $p \in R^n$.

Der Vektor p gibt die Richtung an, in der das oiv-Simplex durchlaufen werden soll. Gesucht ist das maximale Intervall für t derart, daß Δc für alle t aus diesem Intervall oiv ist. Aus Formel (4a) folgt:

$$\sum_{k=1}^{m} t\, \bar a_{ki}\, p_k \leqq \bar c_i + t\, p_i, \qquad i = m + 1(1)\, n.$$

Setzt man

$$\sum_{k=1}^{m} \bar a_{ki}\, p_k - p_i = g_i,$$

so erhält man für t das Intervall

$$s = \operatorname*{Max}_{\substack{m+1 \leqq i \leqq n \\ g_i < 0}} \left\{\frac{\bar c_i}{g_i}\right\} \leqq t \leqq \operatorname*{Min}_{\substack{m+1 \leqq i \leqq n \\ g_i > 0}} \left\{\frac{\bar c_i}{g_i}\right\} = S \tag{16}$$

(Ist $p = e_k$, $1 \leqq k \leqq m$, so ist $g_i = \bar a_{ki}$ und es folgt (6)).

Sei $\Delta\overset{(0)}{c} = \theta$, $\Delta\overset{(i)}{c} = \overset{(i)}{p}$, $i = 1(1)\,r$. Für die Parameter t_i, die nach (16) bei eindimensionaler parametrischer Optimierung bestimmt sind, möge gelten $s_i \leqq t_i \leqq S_i$.

Satz 3 zeigt unmittelbar, daß alle Vektoren

$$\sum_{i=1}^{r} \tilde t_i \overset{(i)}{p} = \text{oiv}$$

sind, wenn

$$\mu_i s_i \leqq \tilde t_i \leqq \mu_i S_i, \qquad i = 1(1)\,r, \quad \sum_{i=1}^{r} \mu_i = 1, \quad \mu_i \geqq 0.$$

Für $\mu_i = 1/r$ besagt dieses Ergebnis, daß das aus den Intervallen $[1/r\, s_i, 1/r\, S_i]\, \overset{(i)}{p}$ gebildete Parallelepiped im oiv-Simplex enthalten ist. Die Überlegungen von Beispiel 2 sind auch hier analog übertragbar.

Beispiel 4:

Ist $K_i K_j \neq 0$, K_i und K_j nichtnegativ, so sei

$$K_{ij} = K_i \,\square\, K_j := \frac{K_i K_j}{K_i + K_j}. \tag{17}$$

Setzt man $K_i = \operatorname{Min} \{|d_i|, D_i\}$, $K_j = \operatorname{Min} \{|d_j|, D_j\}$, so sind die Vektoren $\Delta c_i e_i$ und $\Delta c_j e_j$ oiv, für $|\Delta c_j| \leqq K_j$ and $|\Delta c_i| \leqq K_i$.

Demnach ist $\Delta c_i e_i + \Delta c_j e_j = \text{oiv}$ für

$$-\mu_i K_i \leqq \Delta c_i \leqq \mu_i K_i$$

$$-\mu_j K_j \leqq \Delta c_j \leqq \mu_j K_j,$$

$$\mu_i + \mu_j = 1, \quad \mu_i \geqq 0, \quad \mu_j \geqq 0.$$

Die Zahlen μ_i und μ_j sollen so bestimmt werden, daß $\mu_i K_i = \mu_j K_j$. Es ergibt sich $\mu_j = \frac{K_i}{K_i + K_j}$, bzw.

$$\mu_j K_j = \frac{K_i K_j}{K_i + K_j} = K_i \square K_j = \mu_i K_i .$$

Die Vektoren $\Delta c_i e_i + \Delta c_j e_j$ sind also oiv für alle Δc_i, Δc_j mit $|\Delta c_i| \leqq K_i \square K_j$, $|\Delta c_j| \leqq K_i \square K_j = K_{ij}$.

Ist $\Delta c_l e_l =$ oiv für $|\Delta c_l| \leqq K_l$, so ist $\Delta c_i e_i + \Delta c_j e_j + \Delta c_l e_l =$ oiv für

$$|\Delta c_\nu| \leqq K_{ij} \square K_l = K_{ijl} , \qquad \nu \in \{i, j, l\} .$$

Auf diese Weise kann man sich leicht eine Tabelle wie folgt berechnen:

c_1	$K_1 = \text{Min}\ \{\lvert d_1 \rvert, D_1\}$	
c_2	$K_2 = \text{Min}\ \{\lvert d_2 \rvert, D_2\}$	$K_1 \square K_2 = K_{1,2}$
c_3	$K_3 = \text{Min}\ \{\lvert d_3 \rvert, D_3\}$	$K_3 \square K_{1,2} = K_{1,2,3}$
.	. .	. .
.	. .	. .
.	. .	. .

Dabei seien die c_i so angeordnet, daß $K_i \geqq K_{i+1}$ gilt.

Die Koeffizienten c_1, c_2, c_3 dürfen z. B. simultan und unabhängig voneinander um maximal $\pm K_{1,2,3}$ verändert werden, ohne daß eine neue Optimallösung entsteht.

Anhand des folgenden Beispiels (aus [1], S. 305) seien einige der bisher abgeleiteten Ergebnisse veranschaulicht, wobei in diesem einfachen Fall auch die geometrischen Zusammenhänge sichtbar werden.

Die Aufgabe lautet:

$$\text{Min}!\ Z = -12 x_1 - 40 x_2 - 20 x_3 - 18 x_4$$
$$4 x_1 + 10 x_2 + 9 x_3 + 7 x_4 \leqq 6000$$
$$x_1 + 40 x_2 + x_3 + 3 x_4 \leqq 4000, \quad x_i \geqq 0.$$

Als Optimaltableau erhält man (x_5, x_6 Schlupfvariable):

x_1	x_2	x_3	x_4	x_5	x_6	
1		$\frac{7}{3}$	$\frac{5}{3}$	$\frac{4}{15}$	$-\frac{1}{15}$	$\frac{4}{3} \cdot 1000$
	1	$-\frac{1}{30}$	$\frac{1}{30}$	$-\frac{1}{150}$	$\frac{2}{75}$	$\frac{1}{15} \cdot 1000$
		$\frac{20}{3}$	$\frac{10}{3}$	$\frac{44}{15}$	$\frac{4}{15}$	$\frac{56}{3} \cdot 1000$

Das oiv-Simplex für Störungen von c_1 und c_2 wird nach (4a) durch die folgenden Ungleichungen beschrieben:

$$\frac{7}{3} \Delta c_1 - \frac{1}{30} \Delta c_2 \leqq \frac{20}{3}, \qquad \frac{5}{3} \Delta c_1 + \frac{1}{30} \Delta c_2 \leqq \frac{10}{3},$$

$$\frac{4}{15} \Delta c_1 - \frac{1}{150} \Delta c_2 \leqq \frac{44}{15}, \qquad -\frac{1}{15} \Delta c_1 + \frac{2}{75} \Delta c_2 \leqq \frac{4}{15}.$$

Die Abbildung zeigt das oiv-Simplex im Δc_1, Δc_2-Koordinatensystem sowie das durch die Achsenabschnitte erzeugte Teilsimplex gemäß Formel (11) bzw. (12).

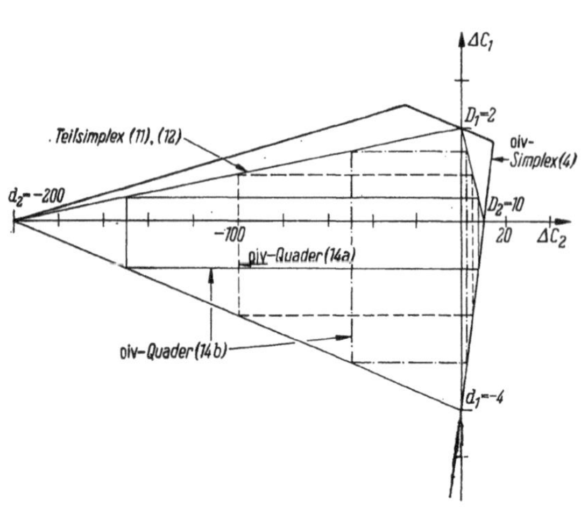

Es ist nach (6):

$$d_1 = -4, \qquad D_1 = 2$$
$$d_2 = -200, \qquad D_2 = 10 .$$

Nach Satz 3 und Formel (14a) sind somit alle Störungsvektoren $\Delta c = \begin{pmatrix} \Delta c_1 \\ \Delta c_2 \end{pmatrix}$ mit

$$-2 \leqq \Delta c_1 \leqq 1 , \qquad -100 \leqq \Delta c_2 \leqq 5 \qquad \text{oiv} .$$

Für $t = 2$ liefert (14b)

$$-3 \leqq \Delta c_1 \leqq \frac{3}{2}, \qquad -50 \leqq \Delta c_2 \leqq \frac{5}{2}$$

oder

$$-1 \leqq \Delta c_1 \leqq \frac{1}{2}, \qquad -150 \leqq \Delta c_2 \leqq \frac{15}{2}.$$

Nach (15) ist der Bereich

$$-\frac{80}{21} \leqq \Delta c_1 \leqq \frac{40}{21}, \qquad -\frac{200}{21} \leqq \Delta c_2 \leqq \frac{10}{21} \qquad \text{oiv} .$$

Nach (7) erhält man für das maximale K, für das $\Delta c = \begin{pmatrix} \Delta c_1 \\ \Delta c_2 \end{pmatrix}$ mit $|\Delta c_i| \leqq K$ oiv ist, den Wert

$$K = \frac{100}{30} : \frac{51}{30} = 1.96 .$$

Nach Formel (17) ergibt sich ein etwas kleinerer Wert, nämlich:

$$K_{1,2} = \frac{2 \cdot 10}{2 + 10} = 1.66 .$$

Werden die Zielfunktionskoeffizienten $c_1, \ldots, c_4$ gestört, so liefert Formel (14a) den 4-dimensionalen oiv-Bereich:

$$-1 \leqq \Delta c_1 \leqq 0.5 , \qquad -50 \leqq \Delta c_2 \leqq 2.5 ,$$
$$-1.66 \leqq \Delta c_3 , \qquad -0.83 \leqq \Delta c_4 .$$

Literatur

1 Dantzig, G. B., Lineare Programmierung und Erweiterungen. Springer Verl., Berlin, Heidelberg, New York 1966.
2 Gass, S. I., Linear Programming, Mc Graw-Hill Book Company.
3 Joksch, H. C., Lineares Programmieren, J. C. B. Mohr (Paul Siebeck), Tübingen 1965.

Eingereicht am 7. 2. 1973

Anschrift: Dr. Volkmar Scharf, 3501 Weimar b. Kassel, Am Eichhölzchen 26 BRD

ZAMM 54, 62 – 66 (1974)

W. H. Isay/H. Niermann

Der in seinen Spitzenwirbeln von der Wasseroberfläche her belüftete Tragflügel*)

Herrn Professor Dr. J. Weissinger zum 60. Geburtstag am 12. Mai 1973 gewidmet.

1. Einleitung

Experimente mit nahe der Wasseroberfläche arbeitenden Propellern zeigen insbesondere bei hohen Froudeschen Zahlen, daß in die Unterdruckgebiete im Kern freier Spitzenwirbel Luft von der Wasseroberfläche hineingesaugt wird. Diese Belüftung führt zu einem erheblichen Absinken des Propellerschubes bzw. des Flügelauftriebs [1], [2], [3]. Es erscheint daher sinnvoll, auch eine theoretische Behandlung belüfteter Spitzenwirbel vorzunehmen. In Anbetracht der Tatsache, daß die eigentliche Propellerströmung recht kompliziert ist, dürfte es zweckmäßig sein, eine Theorie zunächst für den einfacheren Fall des Tragflügels zu entwickeln; diese kann bei Bewährung dann als Modell für die Behandlung des Propellerproblems dienen.

*) Aus dem Institut für Schiffbau der Universität Hamburg.

Wir gehen aus von dem bekannten Strömungsfeld (Potential Φ_Γ) eines Tragflügels im Abstand y_0 von der Wasseroberfläche (Bild 1). Der Einfluß der letzteren läßt sich bei den hier vorausgesetzten hohen FROUDEschen Zahlen durch Spiegelung erfassen. Im Kernbereich der freien Spitzenwirbel herrscht zunächst ein erheblicher Unterdruck, der oft auch mit Kavitationsbildung verbunden ist. Kommt es nun zur Belüftung, so steigt der Druck in den Spitzenwirbeln auf den Luftdruck p_0 der freien Wasseroberfläche an. Wir verzichten hier auf die Behandlung des stark instationären Lufteinbruchsvorganges und beschränken uns auf die Analyse des als stationär vorausgesetzten Zustandes bereits eingetretener Belüftung.

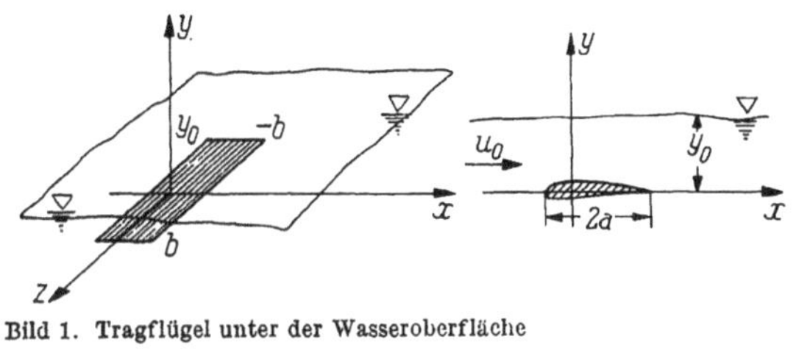

Bild 1. Tragflügel unter der Wasseroberfläche

Die Aufgabe besteht dann darin, ein zusätzliches Geschwindigkeitsfeld (Potential Φ^*) so zu bestimmen, daß im Kernbereich der Spitzenwirbel der Luftdruck p_0 angenommen wird. Da es bei der Ermittlung des den Einfluß der Belüftung enthaltenden Zusatzpotentials Φ^* vor allem auf das Strömungsfeld eines aufgerollten Spitzenwirbels ankommt, erscheint es ausreichend, den Tragflügel im Rahmen dieser Rechnung durch einen einfachen Hufeisenwirbel der Zirkulation $\Gamma_0 =$ konst. zu ersetzen (Bild 2).[1])

Nachdem Φ^* bestimmt ist, hat man aus der normalen Strömungsrandbedingung der erweiterten Traglinientheorie die Flügelzirkulation Γ unter Berücksichtigung der Belüftung zu berechnen.

Das Geschwindigkeitspotential eines Tragflügels der Spannweite $2b$ (als Hufeisenwirbel) im Abstand y_0 von der Wasseroberfläche ist bei hohen FROUDE-Zahlen gegeben durch (Bild 1):

$$\Phi_{\Gamma_0} = \Phi_{(0)}^{\Gamma_0} + \Phi_{\Gamma_0}^{(1)} = \frac{\Gamma_0}{4\pi}\int_{-b}^{b}\int_{0}^{\infty}\frac{y\,dX\,d\zeta}{\sqrt{(x-X)^2+y^2+(z-\zeta)^2}^{\,3}} +$$

$$+\frac{\Gamma_0}{4\pi}\int_{-b}^{b}\int_{0}^{\infty}\frac{(y-2y_0)\,dX\,d\zeta}{\sqrt{(x-X)^2+(y-2y_0)^2+(z-\zeta)^2}^{\,3}}\,. \quad (1)$$

Aus Formel (1) folgt:

$$\frac{\partial\Phi_{\Gamma_0}^{(0)}}{\partial x} = \frac{\Gamma_0}{4\pi}\frac{y}{x^2+y^2}\left(\frac{b-z}{\sqrt{x^2+y^2+(b-z)^2}}+\frac{b+z}{\sqrt{x^2+y^2+(b+z)^2}}\right);$$

$$\frac{\partial\Phi_{\Gamma_0}^{(0)}}{\partial y} = -\frac{\Gamma_0}{4\pi}\frac{x}{x^2+y^2}\left(\frac{b-z}{\sqrt{x^2+y^2+(b-z)^2}}+\frac{b+z}{\sqrt{x^2+y^2+(b+z)^2}}\right)-$$

$$-\frac{\Gamma_0}{4\pi}\left[\frac{b-z}{y^2+(b-z)^2}\left(1+\frac{x}{\sqrt{x^2+y^2+(b-z)^2}}\right)+\right.$$

$$\left.+\frac{b+z}{y^2+(b+z)^2}\left(1+\frac{x}{\sqrt{x^2+y^2+(b+z)^2}}\right)\right];$$

$$\frac{\partial\Phi_{\Gamma_0}^{(0)}}{\partial z} = -\frac{\Gamma_0}{4\pi}\left[\frac{y}{y^2+(b-z)^2}\left(1+\frac{x}{\sqrt{x^2+y^2+(b-z)^2}}\right)-\right.$$

$$\left.-\frac{y}{y^2+(b+z)^2}\left(1+\frac{x}{\sqrt{x^2+y^2+(b+z)^2}}\right)\right]. \quad (2)$$

Analoge Ausdrücke, bei denen lediglich y durch $y - 2y_0$ zu ersetzen ist, ergeben sich für die Ableitungen des gespiegelten Potentials $\Phi_{\Gamma_0}^{(1)}$.

[1]) Γ_0 ist die über Spannweite gemittelte Flügelzirkulation Gl. (20).

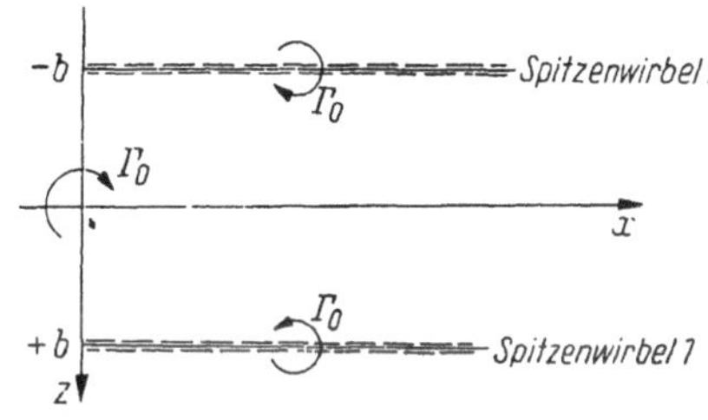

Bild 2. Tragflügel als Hufeisenwirbel

Wir wollen nun voraussetzen, daß die beiden Spitzenwirbel 1 (bei $z = b$) und 2 (bei $z = -b$) des Tragflügels (vgl. Bild 2) so weit auseinander liegen, daß das durch die Belüftung bedingte Zusatzfeld Φ^* in zwei Anteilen Φ_1^* und Φ_2^* unabhängig voneinander bestimmt werden kann. Für die weitere Rechnung ist es dann zweckmäßig, mit der Substitution

$$y = r_1\cos\varphi_1\,, \qquad z = b + r_1\sin\varphi_1 \quad (3)$$

zu einem Zylinderkoordinatensystem überzugehen, dessen Mittelpunkt im Ursprung des Spitzenwirbels 1 (bei $z = b$) liegt.[2]) Da wir uns nun nur für die induzierten Geschwindigkeiten im Kernbereich (also für kleine r_1-Werte) interessieren, können wir noch voraussetzen, daß

$$r_1^2 \ll 4\,b^2\,, \qquad y_0^2 \ll 4\,b^2 \quad (4)$$

ist. Mit dieser Annahme erhält man vereinfachte Ausdrücke für die Geschwindigkeitskomponenten (2).

2. Das durch die Belüftung bedingte Zusatzpotential

Für das infolge der Belüftung des Spitzenwirbels 1 auftretende Zusatzpotential Φ_1^* machen wir den der LAPLACEschen Gleichung genügenden Ansatz:[3])

$$\Phi_1^* = \sum_{\lambda=0}^{z\lambda}\int_0^\infty e^{-\mu x}\,I_\lambda(\mu r_1)\,[\Psi_\lambda^{11}(\mu)\cos\lambda\,\varphi_1 + \Psi_\lambda^{12}(\mu)\sin\lambda\,\varphi_1]\,d\mu \quad (5)$$

Φ_1^* ist aus der Forderung zu bestimmen, daß im Innern des Spitzenwirbels 1 der Luftdruck p_0 angenommen wird.[3])

Mit u_0 als der Anströmgeschwindigkeit und unter Vernachlässigung des Schwerkrafteinflusses führt dieses nach der BERNOULLIschen Gleichung auf die Relation

$$u_0^2 = \left(u_0 + \frac{\partial\Phi_{\Gamma_0}}{\partial x} + \frac{\partial\Phi_1^*}{\partial x}\right)^2 + \frac{1}{r_1^2}\left(\frac{\partial\Phi_{\Gamma_0}}{\partial\varphi_1} + \frac{\partial\Phi_1^*}{\partial\varphi_1}\right)^2 +$$

$$+\left(\frac{\partial\Phi_{\Gamma_0}}{\partial r} + \frac{\partial\Phi_1^*}{\partial r}\right)^2. \quad (6)$$

Es bedarf nun einer gewissen Überlegung, in welchem Bereich Gl. (6) zu erfüllen ist. Wir bedenken, daß hier eine potentialtheoretische Analyse verwendet wird, der im Mittelpunkt $r_1 = 0$ des Wirbels keine echte Bedeutung mehr zukommt. In der wirklichen realen Strömung hat ein Spitzenwirbel bekanntlich einen endlichen Kernradius $R_1(x)$. Nur außerhalb dieses Radius für $r_1 > R_1$ kann in guter Näherung Potentialströmung angenommen werden. Es erscheint daher realistisch, Gl. (6) für $r_1 = R_1$ zu erfüllen und damit den ganzen Bereich $r_1 \leqq R_1$ als belüftet anzunehmen. wie dieses auch in Versuchen [1] beobachtet wurde (vgl. Bild 3). Die genaue Form des Kernradius entzieht sich insbesondere bei turbulenter Strömung einer theoretischen Berechnung und muß experimentell bestimmt werden. Messungen und vereinfachte (teilweise von der Voraussetzung laminarer Strö-

[2]) Die Rechnung für den Spitzenwirbel 2 gestaltet sich ganz analog, so daß hier auf deren Wiedergabe verzichtet werden kann.

[3]) Zur Erläuterung des Ansatzes (5) bemerken wir: Die allgemeinen Lösungen, die für die LAPLACEsche Gleichung bei einem Produktansatz in Zylinderkoordinaten erhalten werden, sind bekanntlich

$$\sim e^{\pm\mu x}\,e^{\pm i\lambda\varphi}\begin{cases} I_\lambda(\mu r) & \lambda \text{ ganz, positiv} \\ N_\lambda(\mu r) & \mu \text{ beliebig}\end{cases}$$

mit I_λ als BESSELscher und N_λ als NEUMANNscher Funktion. Da wir ein im Bereich $x \geqq 0$ reguläres Potential suchen, scheidet $N_\lambda(\mu r)$ wegen seiner Singularität für $\mu = 0$ oder $r = 0$ aus; ebenso ist $e^{\mu x}$ für $\mu > 0$ ungeeignet.

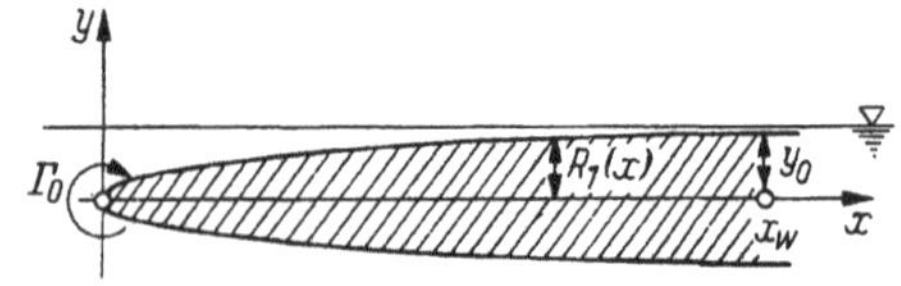

Bild 3. Belüfteter Kernbereich des Spitzenwirbels

mung ausgehende) Rechnungen haben gezeigt, daß der Kernradius mit dem Abstand x vom Flügel zunimmt [4]. Aus tragflächentheoretischen Untersuchungen (z. B. an Propellerflügeln [3]) ist bekannt, daß etwa in der Umgebung der 3/4-Linie des Flügels, also zwischen Flügelmitte und Hinterkante, die niedrigsten Drücke im Spitzenwirbel auftreten. Wir wollen daher den Wert $R_1(a)$, also den Kernradius im Punkt 3/4-Profiltiefe $x = a$ (a halbe Flügeltiefe), aus folgender Betrachtung bestimmen:

Der Druck p_i im Innern $r_1 = 0$ des Kerns eines realen Tragflügelspitzenwirbels mit dem Kernradius $R_1(x)$ und der Zirkulation Γ_0 ist gegeben durch die Relation [4]

$$p_i = p_0 - \varrho \frac{\ln 2}{4\pi^2} \frac{\Gamma_0^2}{R_1^2(x)}, \tag{7}$$

mit p_0 als dem Druck der ungestörten Zuströmung ($p_0 = 1$ kp/cm²). Wenn nun bei $x = a$ ein sehr niedriger Druck (etwa Dampfdruck) für p_i vorliegt, so kann dort $p_0 - p_i \approx p_0$ gesetzt werden, und wir erhalten

$$\frac{1}{b} R_1(a) = \frac{\sqrt{\ln 2}}{2\pi} \sqrt{\frac{\varrho\, u_0^2}{p_0}} \frac{\Gamma_0}{u_0 b} = \frac{\sqrt{\ln 2}}{8} \sqrt{\frac{\varrho\, u_0^2}{p_0}} A_1 . \tag{8}$$

In Formel (8) ist noch entsprechend dem Ansatz (21) die Relation

$$\Gamma_0 = u_0 b \frac{\pi}{4} A_1 \tag{9}$$

verwendet worden.

Die Gleichung (6) ist in ihrer nichtlinearen Gestalt einer theoretischen Analyse nur sehr schwer zugänglich. Es erscheint daher geboten, sie durch einige physikalisch sinnvolle Annahmen zu linearisieren.

Zunächst ist $\frac{\partial \Phi_{\Gamma_0}}{\partial x}$ aus Stetigkeitsgründen im Kernbereich des Spitzenwirbels auf jeden Fall sehr klein, denn an der nahe gelegenen Wasseroberfläche $y = y_0$ wird ja exakt $\frac{\partial \Phi_{\Gamma_0}}{\partial x} = 0$. Wir wollen somit grundsätzlich $\frac{\partial \Phi_{\Gamma_0}}{\partial x}$ gegenüber u_0 vernachlässigen. Im übrigen ist am Kernradius einer Wirbelströmung die Umfangsgeschwindigkeit wesentlich größer als die Radialkomponente. Letzteres geht auch aus Formel (4) hervor. Es liegt daher nahe, mit einer kleinen Größe ε (die nicht näher definiert zu werden braucht) folgende Abschätzungen für die Größenordnung der in Gl. (6) auftretenden Geschwindigkeiten einzuführen:

$$u_1 \sim 1; \frac{1}{r_1} \frac{\partial \Phi_{\Gamma_0}}{\partial \varphi} \sim \sqrt{\varepsilon}; \quad \frac{\partial \Phi_{\Gamma_0}}{\partial r_1} \sim \varepsilon; \quad \operatorname{grad} \Phi_1^* \sim \varepsilon . \tag{10}$$

Unter Vernachlässigung aller Anteile $\sim \varepsilon^2$ geht die Relation (6) in die bezüglich Φ_1^* linearisierte Belüftungsdruckbedingung über:

$$u_0 \frac{\partial \Phi_1^*}{\partial x} + \frac{1}{r_1} \frac{\partial \Phi_{\Gamma_0}}{\partial \varphi_1} \frac{1}{r_1} \frac{\partial \Phi_1^*}{\partial \varphi_1} = -\frac{1}{2} \frac{1}{r_1^2} \left(\frac{\partial \Phi_{\Gamma_0}}{\partial \varphi_1} \right)^2 ,$$

$$(r_1 = R_1(x)) . \tag{11}$$

Im Rahmen der Abschätzung (10) ist es konsequent, bei der Geschwindigkeitskomponente $\frac{1}{r_1} \frac{\partial \Phi_{\Gamma_0}}{\partial \varphi_1}$ nur den wesentlichsten Anteil

$$-\frac{1}{r_1} \frac{\partial \Phi_{\Gamma_0}}{\partial \varphi_1} \approx \frac{\Gamma_0}{4\pi r_1} \left(1 + \frac{x}{\sqrt{x^2 + r_1^2}} \right) \Bigg|_{r_1 = R_1(x)} \tag{12}$$

beizubehalten. Wie sich auf Grund eingehender numerischer Untersuchungen gezeigt hat, spielt auch der zweite Term auf der linken Seite von (11) mit der Größenordnung $\sim \varepsilon^{3/2}$ schon praktisch keine Rolle mehr.

Von dem Ansatz (5) wird dann nur das Glied $\lambda = 0$ benötigt, und aus der Belüftungsdruckbedingung (11) entsteht die Integralgleichung

$$2 u_0 \int_0^\infty e^{-\mu x} I_0(\mu R_1(x))\, \Psi_0^{(11)}(\mu)\, \mu \, d\mu = \frac{\Gamma_0^2}{16\pi^2} \frac{1}{R_1^2(x)} \left(1 + \frac{x}{\sqrt{x^2 + R_1^2}} \right)^2 \tag{13}$$

zur Bestimmung der Funktion $\Psi_0^{(11)}(\mu)$. Für die Lösung von (13) machen wir Gebrauch von der bekannten Integralformel

$$\int_0^\infty e^{-\mu x} I_0(\mu R_1(x))\, d\mu = \frac{1}{\sqrt{x^2 + R_1^2(x)}} . \tag{14}$$

Der Lösungsansatz

$$\Psi_0^{(11)}(\mu) = c_0 \frac{u_0 b}{\mu} \tag{15}$$

mit c_0 als einer zunächst freien Zahlenkonstanten führt dann auf die Relation[4])

$$\frac{2 u_0^2 b}{\sqrt{x^2 + R_1^2(x)}} c_0 = \frac{\Gamma_0^2}{4\pi^2} \frac{1}{R_1^2(x)} . \tag{16}$$

Die Auflösung nach R_1 liefert unter Berücksichtigung von (9)

$$\left(\frac{R_1(x)}{b} \right)^2 = \frac{1}{2} \left(\frac{A_1^2}{128\, c_0} \right)^2 + \sqrt{\frac{1}{4} \left(\frac{A_1^2}{128\, c_0} \right)^4 + \frac{x^2}{b^2} \left(\frac{A_1^2}{128\, c_0} \right)^2} . \tag{17}$$

Die Konstante c_0 wird aus der Bedingung (8) bestimmt; es ergibt sich:

$$c_0 = \frac{1}{2 \ln 2} \frac{p_0}{u_0^2 \varrho} \sqrt{\frac{a^2}{b^2} + \frac{u_0^2 \varrho}{p_0} A_1^2 \frac{\ln 2}{64}} . \tag{18}$$

Der aus Gl. (17) zu entnehmende Kernradius hat in einigem Abstand vom Flügel, für

$$\frac{x^2}{b^2} \gg \left(\frac{1}{2} \frac{A_1^2}{128\, c_0} \right)^2 , \tag{19}$$

die Form

$$R_1(x) \sim \sqrt{x} ,$$

welche aus der Theorie eines Wirbels in zäher Strömung [4] bekannt ist.

Für das durch die Belüftung des Spitzenwirbels 2 (bei $z = -b$) bedingte Potential Φ_2^* wird ein zu Gl. (5) analoger Ansatz mit den Koordinaten x, $y = r_2 \cos \varphi_2$, $z = -b + r_2 \sin \varphi_2$ gemacht. Die gesamte übrige Rechnung bleibt unverändert.

3. Der Einfluß der Belüftung auf die Zirkulationsverteilung und die Kräfte des Tragflügels

Wir betrachten einen Rechtecktragflügel der Tiefe $2a$ und der Spannweite $2b$ (vgl. Bild 1). Liegt die tragende Linie der Zirkulation $\Gamma(\zeta)$ in der z-Achse, so ist die Strömungsrandbedingung der erweiterten Traglinientheorie längs der Geraden $y = 0$, $x = a$, $-b \leqq z \leqq b$ zu erfüllen. Für die durch die Belüftung der Spitzenwirbel bedingten aus den Potentialen Φ_1^*, Φ_2^* folgenden Zusatzgeschwindigkeiten ist dann $r_1 = b - z$, $\varphi_1 = \pi/2$ bzw. $r_2 = b + z$, $\varphi_2 = \pi/2$ zu setzen. Wir bezeichnen ferner mit δ_a den Neigungswinkel der Profilskelettlinie gegenüber der Anströmung u_0 in den Punkten 3/4-Flügeltiefe. Für hohe FROUDEsche Zahlen ergibt sich dann aus der linearisierten Strömungsrandbedingung

[4]) Auf Grund der durchgeführten numerischen Rechnungen hat sich gezeigt, daß praktisch im ganzen x-Bereich (abgesehen von $x = 0$) wegen des sehr kleinen R_1^2-Wertes gesetzt werden kann:

$$1 + \frac{x}{\sqrt{x^2 + R_1^2}} \approx 2 .$$

die Integralgleichung[5])

$$\frac{1}{2\pi}\int_{-b}^{b}\frac{d\Gamma(\zeta)}{d\zeta}\left\{\frac{1}{z-\zeta}+\frac{1}{2}\frac{1}{z-\zeta}\left(\sqrt{1+\left(\frac{z-\zeta}{a}\right)^2}-1\right)+\right.$$

$$+\frac{1}{2}\frac{z-\zeta}{4y_0^2+(z-\zeta)^2}+\frac{1}{\sqrt{a^2+4y_0^2+(z-\zeta)^2}}\times$$

$$\left.\times\left(\frac{1}{2}\frac{a(z-\zeta)}{4y_0^2+(z-\zeta)^2}+\frac{1}{2}\frac{a(z-\zeta)}{4y_0^2+a^2}-\delta_a\frac{y_0(z-\zeta)}{4y_0^2+a^2}\right)\right\}d\zeta$$

$$=-u_0\,\delta_a+u_0\,\delta_a\,c_0\left[\frac{b}{\sqrt{a^2+(b-z)^2}}+\frac{b}{\sqrt{a^2+(b+z)^2}}\right].\tag{20}$$

Gl. (20) ohne die den Belüftungseffekt enthaltenden Anteile (das sind die Glieder mit c_0 auf der rechten Seite) ist aus der Theorie der Unterwassertragflügel seit langem bekannt. Es ist daher naheliegend, die Integralgleichung (20) mit den bekannten Verfahren der Traglinientheorie zu lösen [5], also mit dem Ansatz

$$\Gamma(\zeta)=u_0\,b\sum_{\nu=1}^{z_\nu}A_\nu\sin\nu\,\sigma\,,\qquad \zeta=-b\cos\sigma\,.\tag{21}$$

Dabei ist

$$\Gamma_0=\frac{1}{2b}\int_{-b}^{b}\Gamma(\zeta)\,d\zeta$$

durch die Relation (9) gegeben. Der Vorfaktor c_0 hängt zwar gemäß Gl. (18) in nichtlinearer Form von A_1 ab; jedoch hat sich bei den numerischen Rechnungen gezeigt, daß dieser Einfluß so schwach ist, daß er höchstens in Form einer Iteration berücksichtigt zu werden braucht, also der A_1^2-Term in (18) in erster Näherung vernachlässigt werden darf.

Die Flügelkräfte in x- und y-Richtung (pro Längeneinheit in Spannweitenrichtung) K_x und K_y berechnen wir wie üblich mit dem Satz von Kutta und Joukowski. Dabei verwenden wir für den gespiegelten Flügel das Modell des Hufeisenwirbels (vgl. Gl. (2) für $y - 2y_0$). Es ergibt sich unter Berücksichtigung der bekannten Integralformel

$$\int_0^\infty I_0(\mu(b+z))\,d\mu=\frac{1}{b+z}\tag{22}$$

das Resultat

$$K_x=\varrho\left[\frac{1}{4\pi}\int_{-b}^{b}\frac{d\Gamma}{d\zeta}\frac{d\zeta}{z-\zeta}+\right.$$

$$\left.+\frac{\Gamma_0}{4\pi}\left(\frac{b-z}{4y_0^2+(b-z)^2}+\frac{b+z}{4y_0^2+(b+z)^2}\right)\right]\Gamma(z)\,,\tag{23}$$

$$K_y=\varrho\left[u_0-\frac{\Gamma_0}{8\pi y_0}\left(\frac{b-z}{\sqrt{4y_0^2+(b-z)^2}}+\frac{b+z}{\sqrt{4y_0^2+(b+z)^2}}\right)-\right.$$

$$\left.-u_0\,c_0\left(\frac{b}{b-z}+\frac{b}{b+z}\right)\right]\Gamma(z)\,.\tag{24}$$

Bei der Auswertung der Formel (24) dürfen sinnvollerweise nur z-Werte mit $b-z>R_1$ und $b+z>R_1$, also außerhalb des belüfteten Kernradius liegende Punkte, eingesetzt werden. Abgesehen von der sich aus der Strömungsrandbedingung (20) durch den Einfluß der Belüftung ergebenden Abminderung von $\Gamma(z)$ wirkt auch die in Gl. (24) auftretende Zusatzgeschwindigkeit (Anteil mit c_0) im Sinne einer Verkleinerung des Auftriebs.

[5]) Bei der Berechnung der Belüftungsgeschwindigkeiten ist Gl. (14) zu verwenden; eine y-Komponente tritt wegen $z_A = 0$ in (5) nicht auf.

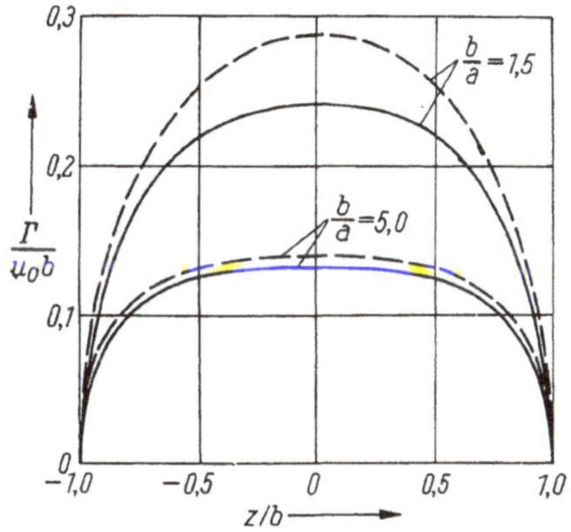

Bild 4. Flügelzirkulation Γ
$\delta_A = 0{,}2$; $y_0/a = 1{,}0$
——— belüftet
– – – unbelüftet

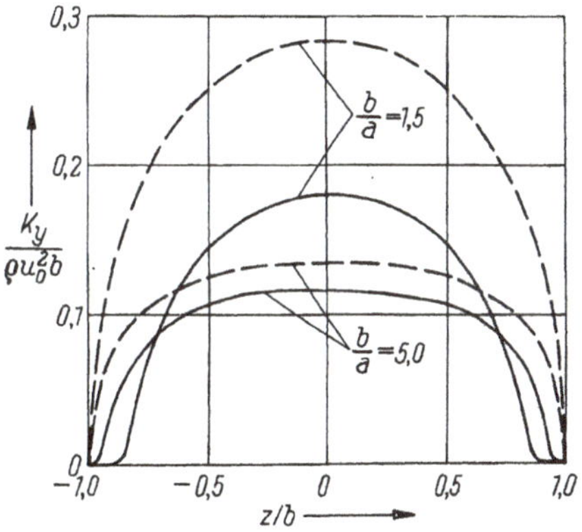

Bild 5. Flügelkraft K_y
$\delta_A = 0{,}2$; $y_0/a = 1{,}0$
——— belüftet
– – – unbelüftet

4. Numerische Ergebnisse

Die vorstehend entwickelte Theorie wurde für Rechteckflügel verschiedener Seitenverhältnisse und eine Reihe von Tauchtiefen und Anstellwinkel ausgewertet. Den Rechnungen liegt der Wert $u_0 = 20$ m/sec, also $\frac{u^2\varrho}{p_0} = 4$ zugrunde; dieser entspricht sowohl der Fahrgeschwindigkeit von Wassertragflügeln (Tragflügelbooten) als auch etwa der Spitzenumfangsgeschwindigkeit von Modellpropellern im Kavitationstank.

Bild 4 zeigt den Verlauf der Zirkulationsverteilung über Spannweite bei belüfteten Flügeln; zum Vergleich ist gestrichelt die Zirkulation der entsprechenden unbelüfteten Flügel bei gleichem Abstand von der Wasseroberfläche eingezeichnet. Bild 5 enthält für die gleichen Beispiele analog den Verlauf der Kraft $K_y(z)$. Man erkennt, daß K_y durch den Belüftungseinfluß in der Nähe der Flügelenden bei $z = \pm b$ erheblich, und zwar wesentlich stärker als die Zirkulation, vermindert wird.

In Bild 6 ist der Gesamtauftrieb

$$A=\int_{-b}^{b}K_y(z)\,dz$$

des Flügels bezogen auf den entsprechenden Wert A_u des unbelüfteten Flügels (mit Einfluß der Wasseroberfläche) in Abhängigkeit von Seitenverhältnis b/a und Tauchung y_0/a dargestellt. Während der Belüftungseinfluß mit zunehmendem Seitenverhältnis schwächer wird, (dieses ist auch anschaulich einleuchtend), ist die Tauchung für eine bereits als eingetreten vorausgesetzte Belüftung nur von geringer Bedeutung. Diese letztere Aussage wird verständlich, wenn wir bedenken, daß in der hier vorgelegten Theorie die sicher mit der Tauchung zusammenhängende Frage, ob und unter

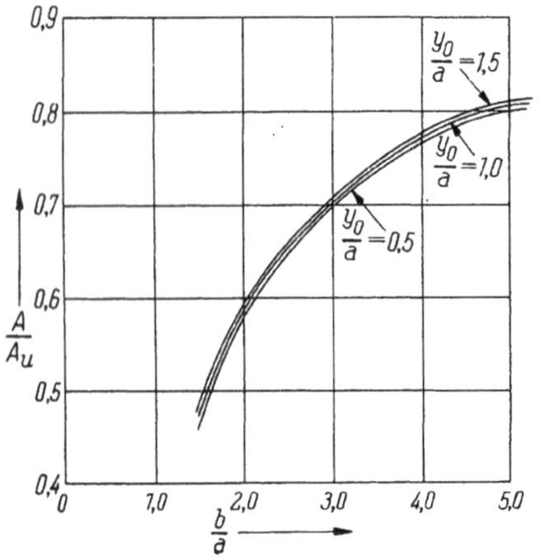

Bild 6. Gesamtauftrieb A bezogen auf den entsprechenden Wert des unbelüfteten Flügels

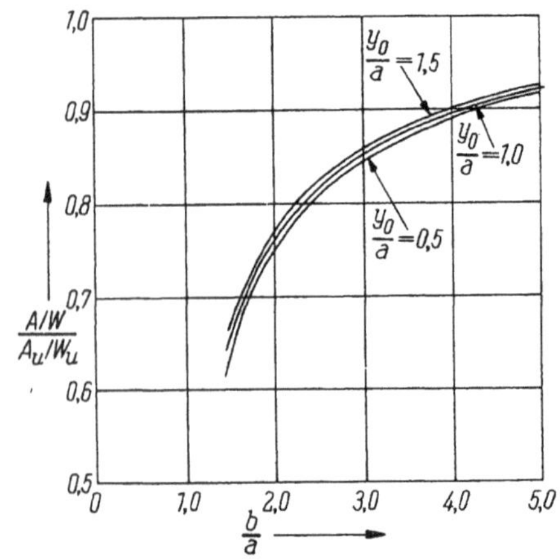

Bild 7. Einfluß der Belüftung auf das Verhältnis A/W (Auftrieb zu Widerstand)

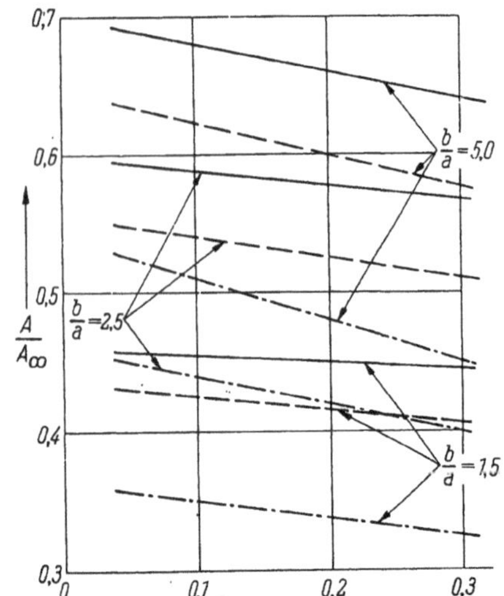

Bild 8. Gesamtauftrieb bezogen auf den entsprechenden Wert im unbegrenzten Medium

——— $y_0/a = 1,5$
– – – $y_0/a = 1,0$
–·–·– $y_0/a = 0,5$

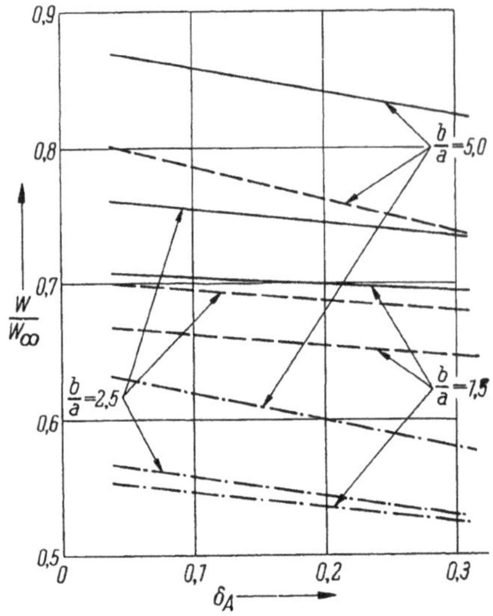

Bild 9. Gesamtwiderstand bezogen auf den entsprechenden Wert im unbegrenzten Medium

——— $y_0/a = 1,5$
– – – $y_0/a = 1,0$
–·–·– $y_0/a = 0,5$

welchen Umständen eine Belüftung eintritt, überhaupt nicht behandelt wird (vgl. Abschnitt 1). Wir bezeichnen ferner mit

$$W = \int_{-b}^{b} K_x(z)\, dz$$

den Gesamtwiderstand des Flügels; entsprechend ist W_u der Wert für den unbelüfteten Flügel mit freier Wasseroberfläche. Bild 7 enthält den Verlauf des Verhältnisses

$$\frac{A}{W} \Big/ \frac{A_u}{W_u}$$

in Abhängigkeit von Seitenverhältnis und Tauchung. Man erkennt, daß dieser Wert stets kleiner als 1 ist; er kann als ein Maß für den Energieverlust bzw. Wirkungsgradverlust durch die Belüftung angesehen werden.

In den Bildern 8 und 9 sind Gesamtauftrieb A und Gesamtwiderstand W des belüfteten Flügels bezogen auf die jeweiligen Werte A_∞ und W_∞ im unbegrenzten Medium dargestellt, und zwar in Abhängigkeit vom Seitenverhältnis b/a, der Tauchung y_0/a und dem Anstellwinkel δ_a. Die Ergebnisse zeigen zumindest qualitativ ein ähnliches Verhalten wie die Resultate von Kraftmessungen an voll belüfteten Propellern [1], [2]. Man erkennt, daß Flügel mit kleinem Seitenverhältnis durch die Belüftung besonders stark beeinflußt werden; dieses ist durch die ja auch sonst bekannte[6]) große Bedeutung der Spitzenwirbel auf die Umströmung eines Flügels von kleinem Seitenverhältnis zu erklären. Andererseits wird durch die freie Wasseroberfläche allein (ohne Belüftung) der Flügel mit großem Seitenverhältnis stärker beeinflußt als mit kleinem [5].

Literatur

1 KRUPPA, C., Testing of partially submerged propellers; Report of Cavitation Committee, Proc. 13th Intern. Towing Tank Conference Berlin-Hamburg 1972.
2 GUTSCHE, F., Einfluß der Tauchung auf Schub und Wirkungsgrad von Schiffspropellern. Schiffbauforschung 6 (1967).
3 ISAY, W. H., Moderne Probleme der Propellertheorie; Kapitel IIC und IB. Springer-Verlag, Berlin-Heidelberg-New York 1970.
4 NEWMAN, B. C., Flow in a viscous trailing vortex. The Aeronautical Quarterly 10 (1959).
5 ISAY, W. H., Propellertheorie, Hydrodynamische Probleme; Kapitel VA; Springer-Verlag, Berlin-Göttingen-Heidelberg 1964.

Eingereicht am 6. 3. 1973

Anschrift: Prof. Dr.-Ing. W. H. ISAY und Dipl.-Ing. H. NIERMANN, Institut für Schiffbau, Universität Hamburg, Hamburg 33, Lämmersieth 90, BRD

[6]) Man denke an die Unterschiede zwischen einfacher und erweiterte sowie linearisierter und nicht-linearisierter Traglinientheorie.

ZAMM 54, 67 (1974)

K. R. Schneider/M. Stefek

Bemerkungen zum Verfahren der zweidimensionalen Sturmschen Kette

Das Problem, die Anzahl der reellen Kurvenäste einer algebraischen Kurve $\mathfrak{K} = \{(x, y) : F(x, y) = 0\}$ in einem vorgegebenen beschränkten Gebiet G zu bestimmen, wird man im allgemeinen durch numerische Ermittlung der Kurvenäste zu lösen versuchen. Dabei kann es leicht vorkommen, daß gewisse Kurvenäste nicht erfaßt werden. In [1] wurde ein Verfahren angegeben — es kann als Verfahren der zweidimensionalen Sturmschen Kette bezeichnet werden —, mit dem nachgeprüft werden kann, ob bei der numerischen Ermittlung der Kurvenäste auch alle Kurvenäste erfaßt wurden. Im wesentlichen wird bei diesem Verfahren der Verlauf einer stückweise konstanten Funktion $N(y)$ mit endlich vielen Sprungstellen in einem Intervall I_y exakt bestimmt.

Betrachten wir in der Funktion $F(x, y)$ y als einen Parameter, so können wir zu $F(x, y)$ als Funktion von x eine Sturmsche Kette $K = \{S_1(x, y), S_2(x, y), \ldots, S_l(x, y)\}$ angeben, falls y gewissen Intervallen angehört. Die Arbeit [1] enthält nun den Schluß, daß das letzte (nicht identisch verschwindende) Glied der Kette K eine Funktion von y allein darstellt, d. h. $S_l(x, y) \equiv \bar{S}_l(y)$. Wie man sich leicht überlegen kann, gilt dieser Schluß aber nur dann, wenn die Kurve $\mathfrak{K}$ keinen mehrfach durchlaufenen Kurvenast besitzt. Im folgenden soll dargelegt werden, wie das in [1] angegebene Verfahren modifiziert werden muß bzw. welche zusätzlichen Überlegungen nötig sind, um die Anwendung des in [1] angegebenen Verfahrens zu sichern, falls $\mathfrak{K}$ mehrfach durchlaufene Zweige besitzt, d. h., falls $S_l(x, y)$ von beiden Variablen abhängt.

Wir bezeichnen mit U die Menge derjenigen y-Werte, für die die Kurve $\mathfrak{K}$ in G singuläre Punkte oder Horizontaltangenten besitzt. Eines der wesentlichsten Ziele des in [1] dargelegten Verfahrens besteht darin, die Menge U bzw. eine Obermenge für U zu ermitteln.

Unter der Voraussetzung, daß $S_l(x, y)$ von x und y abhängt, schreiben wir diese Funktion in der Gestalt $S_l(x, y) \equiv P(y)\, S_l^*(x, y)$, wobei sich von $S_l^*(x, y)$ kein weiterer nur von y abhängiger Faktor abspalten lasse. $S_l^*(x, y)$ ist folglich Teiler von $F(x, y)$, d. h., es gilt $F(x, y) \equiv S_l^*(x, y)\, G(x, y)$. Die Gleichung $G(x, y) = 0$ stellt also ebenfalls die Kurve $\mathfrak{K}$ dar, jedoch mit einfach durchlaufenen Kurvenästen.

Die Glieder $S_r(x, y)$ $(r = 1(1)l)$ der Sturmschen Kette K können wir o. B. d. A. als ganze rationale Funktionen von x und y voraussetzen. Wir bezeichnen den Koeffizienten vor der höchsten x-Potenz in $S_r(x, y)$ mit $A_r(y)$. Dann sei Y die Menge der in I_y gelegenen Nullstellen der Polynome $A_r(y)$ für $r = 2(1)\, l$. Mit Y^* bezeichnen wir die Menge derjenigen y-Werte, für die das Gleichungssystem $G(x, y) = 0$, $G_x(x, y) = 0$ Lösungen in G besitzt. Offensichtlich gilt $U \subset Y^*$. Daraus erhalten wir unmittelbar die folgende Variante des in [1] dargelegten Verfahrens.

Variante 1: Wir berechnen $G(x, y) = F(x, y)/S_l^*(x, y)$ und wenden das in [1] beschriebene Verfahren auf $G(x, y)$ an. Die Ermittlung der neuen Sturmschen Kette erfordert zunächst zusätzlichen Aufwand, der sich jedoch infolge der Verringerung der Grade der beteiligten Polynome lohnt, falls die Kette hinreichend oft zur Berechnung der Anzahl von Nullstellen benutzt wird.

Es läßt sich leicht zeigen, daß $Y^* \subset Y$ gilt. Darauf beruhen die folgenden beiden Varianten.

Variante 2: Wir ermitteln Y und dividieren jedes Glied der Kette K durch $S_l^*(x, y)$ und erhalten so eine neue, zu K äquivalente (vgl. [2], S. 90/91) Sturmsche Kette

$$\tilde{K} = \{\tilde{S}_1(x, y) \equiv G(x, y),\ \tilde{S}_2(x, y) \equiv F_x(x, y)/S_l^*(x, y), \ldots,$$

$$\tilde{S}_l(x, y) \equiv P(y)\},$$

mit der analog zur Kette K weiter verfahren werden kann. Die zugehörige Folge (6) in [1] besteht dann aus der Menge Y und den Nullstellen von $G(x, y)$ auf den Rändern $x = x_0(y)$ und $x = x_1(y)$ des Gebietes G. Hinsichtlich des Aufwandes gilt dasselbe wie bei Variante 1.

Variante 3: Wir ermitteln Y und arbeiten mit der ursprünglichen Kette K. Für die Folge (6) in [1] gilt dasselbe wie bei Variante 2. Falls Y hinreichend viele Werte enthält, wird der Rechenaufwand bei dieser Variante größer sein als bei den ersten beiden.

Ergänzend soll noch bemerkt werden, daß im Fall $l = 2$ die Menge U leer ist. K bricht nämlich genau dann mit $S_2(x, y)$ ab, wenn $F(x, y) \equiv [a(y)\, x + b(y)]^{s_1}$ ist, d. h., $\mathfrak{K}$ wird durch die gebrochen rationale Funktion $x = -\,b(y)/a(y)$ beschrieben.

Literatur

1 Schneider, K. R., Zweidimensionale Sturmsche Ketten, ZAMM 50, S. 752—573 (1970).

2 Obreschkoff, N., Verteilung und Berechnung der Nullstellen reeller Polynome, Berlin 1963.

Eingereicht am 13. 3. 1973

Anschriften: Dr. K. R. Schneider, Zentralinstitut für Mathematik und Mechanik der AdW der DDR, 108 Berlin, Mohrenstr. 39

M. Stefek, Hochschule für Verkehrswesen „Friedrich List“, Sektion Mathematik, Rechentechnik und Naturwissenschaften, 801 Dresden, Friedrich-List-Platz 1, DDR

BUCHBESPRECHUNGEN

S. G. Michlin. Lehrgang der Mathematischen Physik. (Math. Lehrbücher u. Monographien, Bd. XV). XIV + 475 S. m. 49 Abb. Berlin 1972. Akademie-Verlag. Preis geb. 65,– M.

MICHLIN gibt in diesem durch KUHNERT und PRÖSSDORF vorzüglich und fast frei von Druckfehlern übersetzten Buch eine moderne Darstellung der Theorie der linearen partiellen Differentialgleichungen 2. Ordnung vom elliptischen, parabolischen oder hyperbolischen Typ im R^n. Dabei werden die für symmetrische, positiv definite Operatoren A gültigen Variationsmethoden besonders berücksichtigt. Im Vergleich zu früher erschienenen Büchern MICHLINS wird eine noch größere, bewunderswerte Klarheit durch weitgehende Benutzung funktionalanalytischer Methoden erreicht. – In Teil I werden Hilfsmittel über finite Funktionen (Testfunktionen), Mittelfunktionen und verallgemeinerte Ableitungen bereitgestellt. In Teil II wird eine moderne Darstellung der Elemente der Variationsrechnung gegeben, die über das im folgenden Benutzte hinausgeht. In Teil III werden die Vollstetigkeit von FREDHOLMschen regulären oder schwach singulären Integraloperatoren im L_2 gezeigt sowie die vier FREDHOLMschen Sätze für vollstetige Operatoren in einem BANACH-Raum bewiesen. In Teil IV werden Klassifizierung und Randbedingungen für partielle Differentialgleichungen 2. Ordnung kurz dargestellt. Teil V über elliptische Gleichungen hat mit 165 Seiten den größten Umfang. Hervorzuheben sind hier u. a. die Behandlungen 1. der Variationsmethoden bei positiv definiten Operatoren, 2. nicht positiv definiter bzw. nicht selbstadjungierter Operatoren, 3. potentialtheoretischer Methoden bei der LAPLACE-Gleichung und 4. der elliptischen Randwertaufgabe mit am Rand vorgegebener Richtungsableitung, für die i. allg. nicht der FREDHOLMsche Alternativsatz gilt. In Teil VI werden nicht stationäre Gleichungen vom parabolischen oder hyperbolischen Typ behandelt. Im Teil VII ist das Problem der Existenz, der Eindeutigkeit und der Korrektheit für die im Buch diskutierten Aufgaben ausführlich dargestellt. In Anhängen werden behandelt: elliptische Systeme und das CAUCHY-Problem für hyperbolische Gleichungen von W. M. BABITSCH, allgemeine Differentialoperatoren von W. G. MAZJA und nicht lineare elliptische Gleichungen 2. Ordnung von I. J. BAKELMAN. – Für Aufgaben $A\,u - f = 0$ mit $u \in C^{(2)}$ und symmetrischem, positiv definitem Operator A stellt MICHLIN die verallgemeinerte Lösung u_0 methodisch in den Mittelpunkt, nämlich das durch u_0 realisierte Minimum des Funktionals $(A\,u, u) - 2(f, u)$ im energetischen Raum der die Randbedingungen erfüllenden, absolutstetigen Funktionen u mit quadratisch summierbaren ersten Ableitungen. In Teil II wird u. a. das entsprechend verallgemeinerte Eigenspektrum für symmetrische, positiv definite Operatoren A behandelt. In Teil VI werden die Randwertaufgaben der parabolischen Gleichung ($i = 1$) und der hyperbolischen Gleichung ($i = 2$) auf eine gewöhnliche Anfangswertaufgabe mit der Differentialgleichung $d^iu/dt^i + \mathfrak{A}\,u = f(t)$ zurückgeführt, wo $\mathfrak{A}$ ein elliptischer Operator ist und wo für jedes t der abstrakten Funktion u ein Element eines BANACH-Raums zugeordnet ist. Skalare Multiplikation dieser gewöhnlichen Differentialgleichung mit jeder Eigenfunktion des vollständigen Orthonormalsystems der Eigenelemente von $\mathfrak{A}$ liefert die verallgemeinerte Lösung der ursprünglichen, nicht stationären Aufgabe. Für $u_t - u_{tt} = 0$ und für $u_{xx} - u_{xx} = 0$ wird gezeigt, unter welchen verschärften Forderungen an die Anfangsfunktion daraus eine klassische Lösung entsteht. – Das Buch von MICHLIN ist ein durch Beispiele ergänzter, hervorragender Beitrag zur klassischen und modernen Lehr- und Forschungsliteratur der drei Haupttypen linearer partieller Differentialgleichungen 2. Ordnung, dessen Studium für Studenten höherer Semester und für Fortgeschrittene sehr nützlich ist.

Etzenrot E. ADAMS

H. Feldmann, Einführung in ALGOL 60. (Uni-Text). VIII + 112 S. Braunschweig 1972. Friedr. Vieweg & Sohn. Preis brosch. DM 9,80.

ALGOL 60 hat sich im vergangenen Jahrzehnt als wichtigste Programmiersprache für Lehrzwecke durchgesetzt. Bestimmend dafür ist der klare syntaktische Aufbau von ALGOL 60 in Form kontextfreier Produktionsregeln und die nicht allzu umfangreichen semantischen Regeln. Im vorliegenden Buch werden die Produktionsregeln graphisch dargestellt. Es ist so aufgebaut, daß zunächst ein einfacher Teil der Sprache beschrieben wird, der in den folgenden Kapiteln über Ausdrücke, Felder, Blockstruktur, Ein- und Ausgabe ergänzt wird. Zur Ein- und Ausgabe werden Prozeduren nach einem Vorschlag von KNUTH mit Formatsteuerung verwendet. Hierdurch gibt das Buch eine gute Hilfe beim praktischen Programmieren. Die Darlegungen sind durch Beispiele ergänzt. Ein Verzeichnis von Übungsaufgaben und Anhänge über Eingabedarstellung von Zeichen und die Produktionsregeln in BACKUS-Notation beschließen das Buch. Leider sind die Übungsaufgaben oft in der Aufgabenstellung zu knapp (z. B. „Endspielproblem beim Schach", „Automorphismengruppe einer gegebenen Gruppe"), auch sind keine Lösungsvorschläge angegeben.

Insgesamt ist das Buch übersichtlich gestaltet und durch seine knappe und präzise Schreibweise zum Nachschlagen und als Hilfsmittel in der Programmierungspraxis geeignet.

Berlin K.-H. BACHMANN

Annual Review of Fluid Mechanics, Volume 5, 1973. 443 S. mit Fig. u. Abb. Palo Alto, Calif. 1973. Annual Reviews Inc.

Im einleitenden Artikel berichten J. FLÜGGE-LOTZ und W. FLÜGGE über Erinnerungen an die Arbeit im Göttinger Institut für Strömungsforschung, das LUDWIG PRANDTL leitete. Aus vielem anekdotischem Material wird ein recht lebendiges Bild des Kreises um PRANDTL gezeichnet. Aber auch dieser Bericht erinnert wieder einmal daran, daß es eine ausführliche biographische Würdigung des für die Entwicklung der Strömungsmechanik so bedeutenden Wissenschaftlers PRANDTL noch nicht gibt.

Von den Fachaufsätzen müssen auf dem Grenzschichtgebiet zwei Arbeiten hervorgehoben werden.

K. NICKEL schreibt über PRANDTLs Grenzschichten vom Blickpunkt des Mathematikers und E. A. EICHELBRENNER berichtet über den Stand der Erforschung dreidimensionaler Grenzschichten. Der Autor weist auch auf noch offene Fragen hinsichtlich turbulenter Strömungen hin, er empfiehlt besonders auch die experimentelle Erkundung. Speziell den Meßtechniken sind zwei Arbeiten gewidmet: S. S. PENNER und T. JERSKEY beschäftigen sich mit der Lasermessung zum Feststellen örtlicher Geschwindigkeitskomponenten, Dichten und Temperaturen. Mit der Vorlage sehr instruktiver Farbfotografien berichtet H. WERLÉ, wie über das Beobachten von Streichlinien Rückschlüsse auf die Struktur von Wirbelformationen zu ziehen sind.

Diffusionsprobleme wurden angesprochen in den Aufsätzen H. B. FISCHER über die Dispersion und die turbulente Mischung in Kanälen mit offener Oberfläche.

Das Zusammenwirken von Wirbelbewegungen und Wärmefluß rechnet B. GEBHART nach. Er vergleicht mit interferometrischen Messungen, bei denen gut die Dämpfung oder Anfachung von Dichtewellen sichtbar werden.

Auf dem Gebiet der Gasdynamik liegen folgende Untersuchungen vor: A. K. OPPENHEIM und R. I. SOLOVKHIN Gasdynamische Experimente über Explosionen, G. Y. NIENWLAND, B. M. SPEE Tragflügel für den transonischen Bereich, A. FERRI Mischungsgesteuerte Überschallverbrennung und M. N. KOGAN Molekulare Gasdynamik.

In enger Beziehung zur maschinentechnischen Anwendung stehen die Arbeiten über Unterwassertragflügel für Tragflächenboote von A. J. ACOSTA. Über die Flüssigkeitsmechanik der Lagerschmierung schreiben E. A. SAIBEL und N. A. MACKEN.

Während die klassischen Arbeiten die laminare Strömung in Rechnung setzten, wird hier auch über den Einfluß von Wirbeln und von Turbulenz berichtet, aber auch unkonventionelle Anwendungen werden in die Betrachtung einbezogen, so das Gleiten von Automobilrädern, wenn Fahrzeuge mit hoher Geschwindigkeit in Wasserlachen fahren, oder die „Schmierung" der Gelenke im menschlichen Körper. Weiter gehören in diese Gruppe der angewandten Strömungsmechanik die Untersuchung von HORLOCK und LAKSHMINA-

RAYANA über Sekundärströmung in Turbomaschinen und der instruktive Bericht von J. E. McCune und J. L. Kerrebrock über den von Flugzeugtriebwerken erzeugten Lärm.

Auch dieser neueste Band gibt wieder durch eine Reihe von Aufsätzen, die sämtlich auf einem hohen Niveau wissenschaftlicher Forschung stehen, eine ausgezeichnete Übersicht neuer Ergebnisse.

Dresden W. Albring

A. Scholz/B. Schoeneberg, Einführung in die Zahlentheorie, 5. Aufl. (Sammlung Göschen, Band 5131). 128 S. Berlin/New York 1973. Walter de Gruyter. Preis brosch. DM 9,80.

1973 ist die Scholzsche Zahlenthorie, überarbeitet und herausgegeben von Schoeneberg, in der 5. Auflage erschienen. Sehr klar und präzise geschrieben, behauptet sich das Buch nach wie vor auf dem Büchermarkt und dient den Studenten, die sich einen Einblick in die Grundlagen der Zahlentheorie verschaffen wollen, aber nicht den Ehrgeiz haben, bis an die Probleme der heutigen Forschung herangeführt zu werden, als solide Stütze. Es beinhaltet in seinen 4 Kapiteln, von denen besonders das letzte erwähnenswert ist, die Grundtatsachen über Teilbarkeitseigenschaften, Kongruenzen und Restklassen, quadratische Reste und quadratische Formen. Wir wünschen dem Buch, das auch zum Selbststudium gut geeignet ist, weiterhin viele Freunde.

Dresden M. Hasse

Ed. **R. N. Sen** and **C. Weil,** Statistical Mechanics and Field Theory. (Lectures given at the 1971 Haifa Summer School, with additional contributions from participants at the 1971 Europhysics Conference). VII + 333 S. New York/Jerusalem/London 1972. Halsted Press A Division of J. Wiley & Sons/Israel Universities Press. Preis geb. £ 13.15.

Obiges Buch enthält eine ausgezeichnete Zusammenstellung von Vorlesungen einer Sommerschule und einer „Europhysics Conference" des Jahres 1971 über Quantenfeldtheorie und statistische Mechanik.

An den Anfang sind einige Vorlesungsreihen von Araki über normierte Algebren, genauer, über C^*- und W^*-Algebren, ihre Zustände, Darstellungen und Anwendungen auf die Quantenfeldtheorie gestellt. An sie schließen sich einige recht instruktive Betrachtungen von Borchers über die algebraische Formulierung und Aspekte der Wightmanschen Axiome, um einen besseren Zugang zum sogenannten „nichtlinearen Programm" der Quantenfeldtheorie zu erhalten. Weitere interessante Ausführungen über die Grundlagen der Quantenfeldtheorie sind enthalten in den Vorlesungen von Marcel Guenin über das Haagsche Theorem im Zusammenhang mit noch ungelösten Problemen bezüglich des Vakuumzustandes, von O. Steinmann über den Zusammenhang zwischen den traditionellen Wightman- und „LSZ"-Formulierungen, ihre Unterschiede und Vorzüge und über die Beschreibungsmöglichkeit instabiler Teilchen. Kleinere Beiträge von H. Reeh und A. Joseph über Symmetrien, Ströme, infinitesimale Erzeuger und Lie-Identitäten und von Louis Michel über die Wirkung von kompakten Lieschen Gruppen auf verchiedene Mannigfaltigkeiten gehören eng hierzu.

Der andere Teil über Probleme der statistischen Mechanik enthält Vorlesungen über die Grundlagen und Anwendungen auf diesem Gebiete.

Die Betrachtungen von C. M. Edwards über die Verwendung spezieller Begriffe der normierten Algebren in der statistischen Physik, von A. Grossman über impulsartige Konstanten der Bewegung, von A. Verbeure über die Ergodizität in lösbaren Modellen und von W. Minnink über die allgemeinen Eigenschaften von thermodynamischen Zuständen in einer algebraischen Formulierung beziehen sich auf die Grundlagen, während sich die Ausführungen von C. P. Enz über die „Zweiflüssigkeiten-Aspekte der kondensierten Materie" und von F. Siegert über das Isingmodell auf die Anwendungen beziehen. Besonders ausführlich und originell sind die Vorlesungen von H. E. Stanley über Skaleninvarianz und Universalität. Zusammenfassend kann man feststellen, daß es sich bei diesem Buche um ein gelungenes Werk handelt, das für die mathematisch interessierten Physiker, die auf dem Gebiet der statistischen Mechanik und Quantenfeldtheorie arbeiten, eine wertvolle Fundgrube darstellt.

Dresden P. Möbius

Ed. **H. Haken,** Synergetics. Cooperative Phenomena in Multi-Component Systems. 1. Aufl. 279 S. m. Fig. u. Abb. Stuttgart 1973. B. G. Teubner. Preis brosch. DM 54,—.

Der Eindruck, den der vorliegende Symposiumsbericht auf den Leser macht, kann nicht zutreffender umrissen werden, als es durch den Herausgeber selbst geschieht, wenn er im Vorwort schreibt: „At a first glance the reader of the book might be puzzled by the variety of its topics which range from phase-transition-like phenomena of chemical reactions, lasers and electrical current to biological systems, like neuron networks and membranes, to population dynamics and sociology. When looking more closely at the different subjects the reader will recognize, however, that this book deals with one main problem: the behaviour of systems which are composed of many elements of one or a few kinds. We are sure the reader will be surprised in the same way as the participants of a recent symposium on synergetics, who recognized that such systems have amazingly common features. Though the subsystems (e.g. electrons, cells, human beings) are quite different in nature, their joint action is governed by only a few principles which lead to strikingly similar phenomena. It hardly needs to be mentioned that once such common principles are established, they are of an enormous stimulus and help for future research." Das hier erwähnte „Symposium on Synergetics" hat in der Zeit vom 30. April bis 6. Mai 1972 in Schloß Elmau stattgefunden. Das Buch enthält die 24 Vorträge, die dort von Vertretern verschiedener Wissenschaftsgebiete (Physik, Chemie, Biologie, physikalische Chemie, Biochemie, biologische Kybernetik usw.) gehalten worden sind und deren gemeinsames Anliegen von dem — gleichzeitig als Organisator des Symposiums fungierenden — Herausgeber dieses Berichtes in seinem einleitenden, sehr klaren und informativen Vortrag „Introduction to Synergetics" aufgezeigt wird. Mir scheint, daß hier echte Angewandte Mathematik in dem Sinne getrieben wird, daß mathematische Strukturen ausfindig gemacht und aufgestellt werden, die es möglich machen, verschiedenste außermathematische Zusammenhänge unter einem einheitlichen Gesichtspunkt zu betrachten und erfolgreich zu behandeln.

Die Vorträge, auf die hier nicht einzeln eingegangen werden kann und die nicht einmal einzeln aufgezählt werden können, sind in fünf Gruppen aufgeteilt. Mit „Mathematical and Physical Concepts for Cooperative Systems" beschäftigen sich die fünf Beiträge von F. C. Schlögl, R. Kubo, K. Kawasaki, H. Mori und S. Grossmann. Vier Vorträge von R. Graham, H. Thomas, R. Landauer/J. W. F. Woo und I. Prigogine/R. Lefever behandeln den Themenkomplex „Instabilities and Phase-Transition-Like Phenomena in Physical Systems far from Thermal Equilibrium." Fragen der „Biochemical Kinetics and Population Dynamics" sind Gegenstand der Ausführungen von E. H. Kerner, E. W. Montroll/B. J. West, H. Kuhn und des japanischen Teams N. Ogita/A. Ueda/T. Matsubara/F. Yonezawa/H. Matsuda. „Biological Structures" werden von W. Reichardt, B. Julesz, H. R. Wilson, G. Adams, K. Tomita, H. Fröhlich und M. Wagner betrachtet. Schließlich finden sich noch drei Vorträge von M. Mesarovic, dem Kollektiv B. Chinitz/G. Crampton/L. P. Kadanoff/S. C. Schwartz/J. R. Tucker/H. Weinblatt und von W. Weidlich über „General Structures".

Insgesamt gibt der Bericht einen aufschlußreichen Einblick in einen hochinteressanten, durch gemeinsame Gesichtspunkte zusammengehaltenen Komplex verschiedener naturwissenschaftlicher Disziplinen und einen ebenso interessanten Ausblick auf noch nicht gelöste Probleme.

Dresden H. Heinrich

L. Collatz/J. Albrecht. Aufgaben aus der Angewandten Mathematik II. 141 S. m. 29 Bildern. Berlin 1973. Akademie-Verlag. Preis brosch. 14,80 M.

Das aus zwei Bänden bestehende Buch ist als eine erste Einführung in eine ganze Reihe von verschiedenen Teil-gebieten gedacht, die entsprechend einem noch üblichen Sprachgebrauch zur Angewandten Mathematik gezählt werden. Das Buch ist seinem Charakter nach weniger ein Lehrbuch, sondern mehr eine Aufgabensammlung. Wesentliche Betrachtungsweisen und Methoden werden dem Leser unmittelbar an konkreten Beispielen erläutert, bevor eine Schilderung der allgemeineren Sachverhalte und Zusammenhänge erfolgt. Danach werden weitere, bisweilen sehr originelle Aufgaben zur Einübung und Vertiefung vorgelegt. An mathematischen Kenntnissen wird nur soviel vorausgesetzt, wie sie jedem Mathematik- und auch Ingenieurstudenten bereits nach den Anfangsemestern zur Verfügung stehen. Dadurch wird es möglich, einen sehr breiten Kreis von Lesern anzusprechen. Für weiterführende Ergebnisse wird zuständige Spezialliteratur mitgeteilt.

Der vorliegende zweite Band umfaßt die folgenden vier Gebiete: 1. Differentialgleichungen mit den Abschnitten Lösbare Typen, Numerische Methoden (Einschritt-, Differenzenverfahren) und Reihenentwicklungen mit Fehlerabschätzungen; 2. Optimierung und Ungleichungen mit den Abschnitten Optimierung (viele eingekleidete Aufgaben, welche sich meist graphisch lösen lassen), Ungleichungen und Konvexität; 3. Wahrscheinlichkeitsrechnung und Statistik mit den Abschnitten Wahrscheinlichkeitsrechnung, Kombinatorik, Verteilungen, Ausgleichsrechnung und Statistische Tests; 4. Rechenanlagen und ihre Programmierung mit den Abschnitten Algorithmen und Flußdiagramme, Algol-Programmierung (Auszug aus Algol 60), Schaltungsalgebra (Boolesche Algebra mit nur zwei Elementen) und Zahlendarstellungen.

Wie der erste Band, so kann auch der zweite Band dem großen Kreis der Interessenten, die sich mit den behandelten Gebieten beschäftigen wollen oder müssen, als eine nützliche Hilfe für ein erstes Eindringen sehr empfohlen werden.

Dresden J. W. Schmidt

M. A. Krasnosel'skii et al. Approximate solution of operator equations. XII + 484 S. Groningen 1972. Wolters/Noordhoff Publishing. Preis geb. Dfl. 95.00.

Es ist sehr zu begrüßen, daß M. A. Krasnoselski, der in den letzten Jahrzehnten viele wertvolle Ergebnisse über die Lösbarkeit von Operatorgleichungen mit Hilfe topologischer Methoden erzielt hat, zusammen mit seinen Mitarbeitern im vorliegenden Buch zahlreiche Näherungsmethoden zur Lösung von Operatorgleichungen vorführt. Dazu benutzen die Verf. eine einheitliche funktionalanalytische Darstellungsweise, die nicht nur den bekannten Methoden zugute kommt, sondern auch Anlaß zur Entwicklung neuer Näherungsverfahren gibt. Zwecks Raumersparnis werden hier von den ersten drei Kapiteln nur die Überschriften angegeben, dagegen die beiden letzten Kapitel ausführlicher referiert.

Kap. 1. Sukzessive Approximationen.
Kap. 2. Lineare Gleichungen.
Kap. 3. Gleichungen mit glatten Operatoren.
Kap. 4. Projektionsmethoden.

Zunächst werden einige allgemeine Theoreme über die Projektionsmethoden bewiesen, die die großen Schwierigkeiten der Konvergenzfrage erkennen lassen. Eingehend werden dann die Methoden von Bubnov-Galerkin, Galerkin-Petrov sowie die Galerkin-Methode mit kleinen Störungen behandelt. In § 18 werden für die Eigenwerte und Eigenfunktionen der Gleichung $x = \mu T x$ (T zunächst ein linearer kompakter Operator in einem komplexen Banach-Raum E, später ein linearer kompakter selbstadjungierter Operator in einem Hilbert-Raum H) mit den zuvor entwickelten Verfahren Fehlerabschätzungen hergeleitet. Zum Schluß wird die gestörte Galerkin-Methode auf eine nichtlineare Gleichung $x = T x$ mit einem Fréchet-differenzierbaren Operator T angewandt.

Kap. 5. Kleine Lösungen von Operatorgleichungen.

Zunächst werden in § 20 Sätze über die asymptotische Approximation impliziter Funktionen hergeleitet. Besonders hingewiesen sei auf die Behandlung der progressiv eliminationsfähigen Systeme impliziter Funktionen in § 21 (russisch: грубые системы), wobei noch angenommen wird, daß die letzte Operatorgleichung des gestaffelten Systems lösbar ist. Solche Systeme können stets nur endlich viele Lösungen haben.

Beim Studium der nichtlinearen Operatorgleichung $f(\lambda, x) = 0$ mit $f(\lambda_0, x_0) = 0$ tritt bekanntlich das Problem der Verzweigungslösungen auf. Für ihre Bestimmung lassen sich durch Zerlegung des abgespaltenen linearen Operators in geeignete Paare von Projektionsoperatoren verschiedene Verzweigungsgleichungen aufstellen. Nach Übertragung der bekannten Methoden von M. A. Ljapunov und E. Schmidt auf Operatorgleichungen wird gezeigt, daß die zugehörigen Verzweigungsgleichungen genau dieselben Lösungen besitzen. Wegen der Schwierigkeiten, die mit der Aufstellung der exakten Verzweigungsgleichung verbunden sind, wird in § 22 die Konstruktion asymptotischer Approximationen von Verzweigungsgleichungen sowie die von V. B. Melamed herrührende Methode der formalen Potenzreihen erörtert. Wie die Theoreme über die Lösbarkeit von Operatorgleichungen mit einem analytischen Operator in 22.7 zeigen, können auch hier die Fälle eintreten, daß die Verzweigungsgleichung keine Lösung oder eine isolierte Lösung oder unendlich viele Lösungen hat. Jedoch kommen diese Sonderfälle bei Operatorgleichungen, die aus der Mechanik stammen, nur sehr selten vor (vgl. die Arbeiten von U. Kosel, Wiss. Z. TH Ilmenau **15**, H. 4/5, 53–66 (1969), Beitr. Analysis **3**, 135–145 (1972) und J. Vogel, Wiss. Beitr. 1968/69 (M 1) Univ. Halle, 129–136, Wiss. Z. TH Ilmenau **15** (1969) H. 4/5, 67–75). Zur Behandlung der Verzweigungsgleichung gehen die Verf. anschließend näher auf das gut bekannte System von Nekrasov-Nazarov ein, das i. allg. gewisse Vorteile für die Berechnung der gesuchten Koeffizienten bietet.

In § 24 wird eine nichtlineare Operatorgleichung $A(\lambda, x) = 0$ mit $A(\lambda, \theta) = 0$ in einem reellen Banach-Raum auf Verzweigungspunkte hin untersucht. Als Definition des Verzweigungspunkts wird hier die folgende gebraucht, die auf S. 452 mit einem Druckfehler behaftet ist. Es sei $A(\lambda, x) = 0$ mit $A(\lambda, \theta) = 0$ gegeben. Dann heißt die Zahl λ_0 ein Verzweigungspunkt des Parameters λ, wenn für jedes $\varepsilon > 0$ in der Umgebung $|\lambda - \lambda_0| < \varepsilon$ von λ_0 mindestens eine nichttriviale Lösung dieser Gleichung mit $||x|| < \varepsilon$ existiert.

Trotz seiner theoretischen Zielsetzung kann das Studium dieses Buches auch jedem Mathematiker sehr empfohlen werden, der sich für die numerische Behandlung von Näherungsverfahren interessiert.

Halle H. Schubert

Hrsg. **H.-J. Treder**, Zur Geschichte der Erde und des Kosmos. (AdW, Forschungsbereich Kosmische Physik, Heft 1). 74 S. Akademie-Verlag. Berlin 1973. Preis brosch. 10,– M.

Dieses Heft eröffnet eine neue Schriftenreihe, die über Fortschritte der kosmischen Wissenschaften — von der Geologie bis zur Kosmologie — Auskunft geben soll unter dem besonderen Gesichtspunkt ihres Beitrages zum allgemeinen wissenschaftlichen Weltbild. Es ist nun die Kategorie der Historizität, die diese Wissenschaften den Laboratoriumswissenschaften hinzufügen. Für diese gilt: Unter gleichen Umständen geschieht überall und immer dasselbe. Gleiche Umstände erfordern aber in Strenge die vollständige Aufhebung aller äußeren Einflüsse. Was das angesichts der für die Kosmologie konstitutiven Universalität des Gravitationsfeldes genau bedeutet, untersucht H. J. Treder in seinem Beitrag „Zur Frage der Zeitabhängigkeit der Naturgesetze". Er führt zur Unterscheidung zwischen den ahistorischen, im isolierten Laboratorium gewonnenen und daher lokalen, differentiellen Elementargesetzen einerseits und den bei Aufhebung der Isolation zu Tage tretenden Erscheinungen der Historizität andererseits, die in der Kosmologie letztlich globalen Charakters sind. Insofern globale Gesetze aus den differentiellen durch Integration entstehen, hat man Randbedingungen zu akzeptieren, die bezüglich des realen Kosmos erwiesenermaßen zeitabhängig sind. So erscheinen die zeitlich veränderlichen Randbedingungen als Quelle der kosmischen Historizität, und sie überträgt sich auf alle Systeme die zeitlich und räumlich so ausgedehnt sind, daß sie von kosmischen Strukturen affiziert sind. Aber ist dieser Umstand die einzige Ursache der Historizität?

Es ist ja keineswegs immer die Integration von Differentialgleichungen, die zum Verständnis des Makrosystems führt, wie die statistische Physik mit ihrer — vielleicht nicht Historizität so doch — Irreversibilität zeigt. Somit empfiehlt es sich, den abgegeschlossenen Fundamentalsystemen mit ihren ahistorischen Elementargesetzen, offene Systeme im Hinblick auf ihre komplizierte bzw. komplexe Natur gegenüberzustellen, und von daher „Gedanken zur Frage nach der Historizität der Naturgesetze" zu entwickeln. In einem so überschriebenen Aufsatz tut das A. Watznauer. Neben allgemeinen Erwägungen wird hier speziell — in einem Abschnitt über „die Erde als komplexes, offenes System" — der Standpunkt des Geologen illustriert.

Bei allem erkenntnistheoretischen Reiz der ganzen Fragestellung darf nicht vergessen werden, daß es sich um Naturwissenschaft handelt, und das bedeutet, immer wieder darauf zu rekurrieren, was und in welchem Umfang empirisch gesichert ist. Hierdurch zeichnet sich der Beitrag von H. Lambrecht in einer Weise aus, daß er „zur Kosmogonie der interstellaren Materie" eigentlich erst im letzten Drittel kommt. Zunächst soll der Nachweis erbracht werden, „daß man Kosmogonie treiben kann, ohne in prinzipielle Widersprüche fundamentaler Art zu geraten". Dieser Nachweis hat in erster Linie — und das ist Voraussetzung für naturwissenschaftliches Reden von kosmischer Historizität — die Zeitskalen und — Richtungen im Universum als prinzipiell beobachtbar darzutun. Dementsprechend analysiert Lambrecht hier, mit welcher Sicherheit (oder Unsicherheit) wir über die Zeiten des Beginns der Expansion, der Entstehung der ältesten Sterne unserer Galaxis, der Bildung der schweren Elemente im Geflecht der empirischen und theoretischen Sachverhalte verfügen können. Interessanterweise macht er dabei dieses Geflecht selber zum Gegenstand der historischen Kritik, führt eindrucksvolle Beispiele theoretischer Voreingenommenheit an, und zeigt so, daß über die Geschichte des Kosmos auch über die Geschichte der kosmischen Wissenschaften zu reden heißt.

Dresden H. G. Schöpf

P. Kandzia/H. Langmaack. Informatik: Programmierung. (Teubner Studienbücher). Band 18, 234 S. m. 54 Fig. Stuttgart 1973. B. G. Teubner. Preis brosch. DM 18,80.

Dieses aus Vorlesungen des zweiten Verfassers hervorgegangene Buch gibt eine Einführung in die theoretischen Grundlagen der Programmierung. Hierzu wird auf den in der mathematischen Grundlagenforschung und mathematischen Logik üblichen Begriffsbildungen aufgebaut. Ein erstes Kapitel vermittelt unter der Überschrift „Mathematische Grundlagen der Programmierung" Kenntnisse über Grundbegriffe der Logik und Mengenlehre, über Relationen und Funktionen, Zeichenreihe und Termsysteme. Im zweiten Kapitel wird die Verarbeitung von Zeichenreihen mit Hilfe von Algorithmen betrachtet, speziell wird auf Markoffsche Algorithmen und Turing-Maschinen eingegangen. Das dritte Kapitel ist den eigentlichen Grundbegriffen der Programmierung gewidmet, in ihm wird der Aufbau von Programmen aus Grundelementen betrachtet. Dabei werden zur Darstellung und Erläuterung die Sprachen ALGOL 60, ALGOL 68 und EULER herangezogen. Die mit der Programmierung eng verbundenen Eigenschaften der zu verarbeitenden Daten werden im vierten Kapitel betrachtet. Schließlich enthält das fünfte Kapitel eine Darstellung der Übersetzung von Termen und von kontextfreien Sprachen mit Hilfe der Kellermethode.

Im gesamten Buch wird streng darauf geachtet, eine präzise mathematische Ausdrucksweise zu verwenden. Ein Programm wird (maschinenorientiert) als Abbildung des Maschinenspeichers in sich definiert bzw. bei Verwendung von Programmiersprachen als Abbildung der Menge von Tripeln aus Speicherzellen, Werten und Marken in sich. Hierzu werden elementare Transformationen eingeführt, z. B. Speicherzuweisungen, Wertzuweisungen, Sprünge. Die in den betrachteten Programmiersprachen vorkommenden Anweisungen werden auf derartige elementare Transformationen zurückgeführt. Für rekursive Prozeduren wird gezeigt, daß die zugeordnete Transformation minimaler Fixpunkte des Prozedurkörpers ist. Als Daten werden neben elementaren Daten Felder und strukturierte Daten betrachtet, die in den erwähnten Programmiersprachen vorkommen. Der verwendete Variablenbegriff beruht in Anlehnung an ALGOL 68 auf dem Adressenbegriff, d. h. die Bezeichnungen für Variable werden als Adressen interpretiert. Dementsprechend wird auch die Operation des Dereferenzierens (Übergang von der Adresse zum Wert) benötigt.

Den Schluß des Buches bildet die Beschreibung der sequentiellen Übersetzung von Termsystemen und kontextfreien Sprachen. Die Übersetzung kontextfreier Sprachen basiert auf der syntaktischen Analyse von unten (bottom up) mit Hilfe eines Kellers durch kanonische Reduktion (Linksreduktion). Für eine Sprache mit einer kanonisch reduzierbaren Grammatik der Ordnung k(L $R(k)$-Grammatik) werden hinreichende Bedingungen für die Eindeutigkeit angegeben. Abschließend folgt anhand einer Untersprache von ALGOL 60 die Beschreibung eines Übersetzungsalgorithmus.

Wie aus diesem Inhalt zu ersehen ist, handelt es sich bei diesem Buch nicht um eine erste Einführung in die Programmierung, die für ein Erlernen der Programmierung und seine praktische Anwendung in der elementaren Datenverarbeitung erforderlich ist. Das Buch gibt vielmehr für Interessenten an der Grundlagenforschung auf dem Gebiet der Programmierung und für Systemprogrammierer eine sehr klare Darstellung sowohl der Grundbegriffe als auch wichtiger Einzelheiten, z. B. auf dem Gebiet der automatischen Übersetzung. Durch die knappe und mathematisch präzise Schreibweise ist ein sorgfältiges Durcharbeiten angebracht. In diesem Falle wird der interessierte Leser einen hohen Gewinn haben, da Resultate der modernen Grundlagenforschung in ausgezeichneter Weise aufbereitet dargestellt werden.

Berlin K.-H. Bachmann

R. Kulikowski/G. Wunsch, Optimale und adaptive Prozesse in Regelungssystemen. Band 1. 180 S. m. 43 Abb. Berlin 1973. VEB Verlag Technik. Preis brosch. 12,— M.

Nachdem die klassische Regelungstheorie ihre Bemühungen auf das Problem der „Stabilität" ausgerichtet hatte, ist von der modernen Regelungstheorie der optimale und selbsteinstellende Regelvorgang in den Mittelpunkt gestellt worden. Zu seiner Behandlung ist ein erweitertes mathematisches Handwerkszeug notwendig, das unter anderem auch die moderne Algebra und die Funktionalanalysis umfaßt. Das vorliegende Buch bringt diese Grundlagen, wobei die mathematischen Hilfsmittel im Einzelnen nicht entwickelt, sondern nur angegeben werden. Es behandelt im ersten Abschnitt die zur Darstellung von linearen Signalräumen notwendigen Begriffe wie: Mengen, Maße, abstrakte Räume, Integrale nach Stieltjes und Lebesgue, sowie Grundbegriffe der Wahrscheinlichkeits- und Korrelationstheorie. Der zweite Abschnitt behandelt die linearen Systeme mit ihren zugehörigen Operatoren und Funktionalen. Der dritte Abschnitt schließlich gibt die Methoden zur Behandlung nichtlinearer Systeme an, wobei nichtlineare Operatoren und Funktionale, Iterationsverfahren und Stabilitätsbedingungen gebracht werden.

Der vierte und fünfte Abschnitt der polnischen Originalausgabe behandelt dann schließlich die optimalen und adaptiven Systeme mit ihren Optimalitätskriterien, den Extremaleigenschaften der Funktional- und Variationsprobleme. Diese beiden Abschnitte sind in der deutschen Ausgabe zu einem gesonderten zweiten Band zusammengefaßt. Der hier vorliegende erste Band der deutschen Übersetzung ist dagegen von G. Wunsch noch durch Kapitel ergänzt worden, in denen die Zustandstheorie dynamischer Systeme behandelt wird.

Das wesentliche Schrifttum ist angegeben. Einige bis ins Gerätetechnische vorgetriebene Beispiele stellen eine Verbindung zur Praxis dar. Bilder, Druck und äußere Aufmachung sind gut. Das Buch gibt eine empfehlenswerte, sorgfältig ausgearbeitete Darstellung der mathematischen Grundlagen optimaler und adaptiver Regelvorgänge.

Darmstadt W. Oppelt

Die besprochenen und angezeigten Bücher sind durch den Buchhandel zu beziehen.

EINGEGANGENE BÜCHER

Bei der Schriftleitung sind folgende Bücher eingegangen (ausführliche Besprechung bleibt vorbehalten).

Mangoldt/Knopp, Einführung in die Höhere Mathematik. IV. Band v. F. Lösch. XI + 612 S. m. 9 Fig. Stuttgart/Leipzig 1973. S. Hirzel Verlag. Preis geb DM 38,–.

G. H. Meyer, Initial Value Methods for Boundary Value Problems. (Mathematics in Science and Engineering, Volume 100.) X + 220 S. m. Fig. New York/London 1973. Academic Press. Preis geb. $ 14.50.

R. B. Dingle, Asymptotic Expansions: Their Derivation and Interpretation. XV + 521 S. m. Fig. London/New York 1973. Academic Press. Preis geb. £ 13.50.

I. Schur, Gesammelte Abhandlungen. In 3 Teilbänden. XV, IV + 1465 S. Berlin/Heidelberg/New York 1973. Springer-Verlag. Preis geb. DM 198,–.

Ed. **A. Sawczuk,** Foundations of Plasticita. Papers contributed to the international symposium on Warsaw, August 30 – September 2, 1972. XI + 605 S. m. Fig. Leyden 1973. Noordhoff International Publishing. Preis geb. Dfl. 144,25.

F. Oberhettinger / L. Badii, Tables of Laplace Transforms. VII + 428 S. Berlin/Heidelberg/New York 1973. Springer-Verlag. Preis brosch. DM 39,–.

Blaschke / Leichtweiß, Elementare Differentialgeometrie. 5. vollst. neubearb. Aufl. (Die Grundlehren der math. Wissenschaften, Band 1.) X + 369 S. m. 37 Abb. Berlin/Heidelberg/New York 1973. Springer-Verlag. Preis geb. DM 96,–.

Ch. F. Osgood (Ed.), Diophantine Approximation and Its Applications. (Proceedings of a Conference on Held in Washington D.C., June 6–8, 1972.) IX + 356 S. m. Fig. New York/London 1973. Academic Press. Preis geb. $ 12.50.

K. Reinschke, Zuverlässigkeit von Systemen. Band 1. (Theoretische Grundlagen der Technischen Kybernetik.) 227 S. m. Fig. Berlin 1973. VEB Verlag Technik. Preis brosch. 22,– M.

W. Dück, Numerische Methoden der Wirtschaftsmathematik II. (Math. Lehrbücher u. Monographien, Band XVIII). XIV + 529 S. m. 108 Abb. Berlin 1973. Akademie-Verlag. Preis geb. 42,– M.

Hrsg. **M. A. Arbib,** Algebraische Theorie abstrakter Automaten, formaler Sprachen und Halbgruppen. 332 S. m. 60 Abb. u. 4 Tab. Berlin 1973. Akademie-Verlag. Preis geb. 58,– M.

H. G. Garnir / M. De Wilde et **J. Schmets,** Analyse Fonctionnelle. (Mathematische Reihe, Band 45.) 375 S. Basel/Stuttgart 1973. Birkhäuser Verlag. Preis geb. sfr. 96,–.

Mangoldt / Knopp, Einführung in die Höhere Mathematik. IV. Band v. F. Lösch. XV + 612 S. m. 9 Fig. Leipzig 1973. S. Hirzel Verlag. Preis geb. 28,– M.

E. Scheibe, The Logical Analysis of Quantum Mechanics. (International Series of Monographs in Natural Philosophy, Volume 56.) VII + 204 S. Oxford/New York/Toronto/Sydney/Braunschweig 1973. Pergamon Press. Preis geb. £ 6.50.

M. Tiron, Teoria Erorilor de Masurare. 397 S. m. Fig. Bukarest 1972. Editura Tehnica. Preis geb. Lei 22.

I. Tomescu, Introducere in Combinatorica. 250 S. m. Fig. Bukarest 1972. Editura Tehnica.

D. Moisil u. **G. Moisil,** Teoria si practica elipsometriei. 252 S. m. Fig. Bukarest 1973. Editura Tehnica. Preis brosch. Lei 12.

P. Bradeanu, Mecanica Fluidelor. 543 S. m. Fig. Bukarest 1973. Editura Tehnica. Preis geb. Lei 41.

A. Ripianu, Mecanica solidului rigid. 429 S. m. Fig. Bukarest 1972. Editura Tehnica. Preis geb. Lei 42.

L. L. Helms, Einführung in die Potentialtheorie. 305 S. Berlin/New York 1973. Walter de Gruyter. Preis geb. DM 48,–.

E. S. Wentzel / L. A. Owtscharow, Aufgabensammlung zur Wahrscheinlichkeitsrechnung. (Mathematische Lehrbücher und Monographien, Band 22.) In deutscher Sprache herausgegeben von P. Franken. 354 S. m. 132 Abb. u. 2 Tab. Berlin 1973. Akademie-Verlag. Preis geb. 32,– M.

H. D. Baehr, Thermodynamik. 3. neubearb. Aufl. XII + 440 S. m. 271 Abb. u. zahlr. Tab. Berlin/Heidelberg/New York 1973. Springer-Verlag. Preis geb DM 48.,–.

F. Reutter / E. Heyne, Untersuchungen über die automatische Lösung von Aufgaben der konstruktiven Geometrie. (Forschungsbericht des Landes Nordrhein-Westfalen Nr. 2300.) Opladen 1973. 57 S. m. 10 Abb. Westdeutscher Verlag. Preis brosch. DM 13,60.

J. Casti / R. Kalaba, Imbedding Methods in Applied Mathematics. (Series Applied Mathematics and Computation No. 2.) 306 S. m. Fig. London/Amsterdam/Don Mills/Ontario/Sydney/Tokyo 1973. Addison-Wesley Publishing Company. Preis brosch. $ 8.50.

G. Strang / G. J. Fix, An Analysis of the Finite Element Method. (Series in Automatic Computation. XIV + 306 S. m. Fig. Englewood Cliffs, N.J. 1973. Prentice-Hall, Inc.

J. D. Lambert, Computational Methods in Ordinary Differential Equations. (Introductory Mathematics for Scientists and Engineers.) XV + 278 S. m. Fig. London/New York/Sydney/Toronto 1973. John Wiley & Sons. Preis geb £ 5.50.

F. F. Bonsall / J. Duncan, Numerical Ranges II. (London Mathematical Society Lecture Note Series 10.) VII + 179 S. Cambridge 1973. At the University Press. Preis brosch. £ 2.40.

W. Giloi / H. Liebig, Logischer Entwurf digitaler Systeme (Hochschultext). IX + 307 S. m. 196 Abb. Berlin/Heidelberg/New York 1973. Springer-Verlag. Preis brosch. DM 32,–.

S. Mizohata, The theory of partial differential equations. XI + 490 S. Cambridge 1973. University Press. Preis geb. £ 12.50.

D. Klaua, Grundbegriffe der axiomatischen Mengenlehre (WTB 82 u. 105). Berlin 1973. Akademie-Verlag. Teil 1: Einführung in die Allgemeine Mengenlehre II/1. 142 S. Teil 2: II/1. 193 S. Preis pro Band 8,– M.

Bestellungen sind an das zuständige Postamt oder eine Buchhandlung zu richten, Bestellungen aus dem Ausland an die Firma Buchexport, Volkseigener Außenhandelsbetrieb der DDR, 701 Leipzig, Leninstr. 72, oder direkt an den Akademie-Verlag, DDR-108 Berlin, Leipziger Str. 3–4

Herausgeber und Hauptschriftleiter: Prof. Dr.-Ing. habil. Dr. techn. h.c. H. Heinrich, 8027 Dresden, Friedrich-Hegel-Str. 31. Verlag: Akademie-Verlag, DDR-108 Berlin, Leipziger Straße 3–4; Fernsprecher: 220441, Telex-Nr. 112020, Postscheckkonto: Berlin 35021. Bestellnummer dieses Heftes: 1009/54/1. Die Zeitschrift für Angewandte Mathematik und Mechanik erscheint jährlich mit 12 Heften, einschließlich Tagungsheft. Bezugspreis je Heft 22,– M (Sonderpreis für die DDR 18,– M); Bezugspreis eines Bandes 264,– M zuzüglich Versandkosten (Sonderpreis für die DDR 216,– M). Veröffentlicht unter Lizenznummer 1282 des Presseamtes beim Vorsitzenden des Ministerrates der Deutschen Demokratischen Republik. Gesamtherstellung: VEB Druckerei „Thomas Müntzer", 582 Bad Langensalza. Printed in the German Democratic Republic.

JEAN-PIERRE SERRÈ

Lineare Darstellung endlicher Gruppen

(Übersetzung aus dem Französischen)

In deutscher Sprache herausgegeben von GÜNTHER EISENREICH

1972. VIII, 104 Seiten – 2 Abbildungen – 8° – 14,– M

Bestell-Nr. 761 583 8 (5878)

Der erste relativ elementare Teil des vorliegenden Buches enthält die grundlegenden Sätze der klassischen Theorie der linearen Darstellungen endlicher Gruppen (Begriff der Darstellung und Teildarstellung, Irreduzibilität, Ausreduzieren und Darstellungen Kroneckerprodukt, Schursches Lemma) mit einer ausführlichen Behandlung der Theorie der Gruppencharaktere und einem referierenden Überblick über die Darstellungen kompakter Gruppen. Der zweite Teil enthält vor allem für den Mathematiker interessante Verallgemeinerungen der Theorie auf beliebige Körper der Charakteristik 0, während sich sein dritter Teil, der eine Einführung in die Braunersche Theorie für den Übergang von der Charakteristik 0 zur Charakteristik p und umgekehrt enthält und zu seinem Verständnis speziellere algebraische Kenntnisse erfordert, mehr an den Spezialisten wendet.

Bestellungen durch eine Buchhandlung erbeten

A K A D E M I E - V E R L A G

DDR-108 Berlin, Leipziger Str. 3–4

34115

EBERHARD ZEIDLER

Beiträge zur Theorie und Praxis freier Randwertaufgaben

(Schriftenreihe der Institute für Mathematik der Akademie der Wissenschaften der DDR)

1971. XII, 222 Seiten — 13 Abbildungen — gr. 8° — 46,— M

Bestell-Nr. 761 417 3 (2069/A/9)

Eine umfangreiche Klasse permanenter Wellenbewegungen läßt sich durch konforme Abbildung auf nichtlineare funktionentheoretische Randwertaufgaben zurückführen. Das Ziel der Untersuchung ist es, für diesen Problemkreis unter wesentlicher Benutzung von Methoden der nichtlinearen Funktionalanalysis eine einheitliche Theorie aufzubauen, die von reinen Existenzaussagen bis zur expliziten Konstruktion der Lösungen reicht. Diese Darstellung ist so gehalten, daß sie von Studenten mittlerer Semester verstanden werden kann.

Um dem Leser ein möglichst geschlossenes Bild über diesen Problemkreis zu vermitteln, wird auch auf die historische Entwicklung und die vorhandene Literatur, die bis in die neueste Zeit reicht, ausführlich eingegangen.

Bestellungen durch eine Buchhandlung erbeten

AKADEMIE-VERLAG

DDR-108 Berlin, Leipziger Str. 3—4

Bei Fragen zur Produktsicherheit wenden Sie sich bitte an:
If you have any questions regarding product safety,
please contact:

Walter de Gruyter GmbH
Genthiner Straße 13
10785 Berlin
productsafety@degruyterbrill.com